Trends and Issues in Global Tourism

For further volumes:
http://www.springer.com/series/8406

Roland Conrady • Martin Buck

Editors

Trends and Issues
in Global Tourism 2012

In Collaboration with Pia Viehl

 Springer

Editors
Prof. Dr. Roland Conrady
Worms University of Applied Sciences
Department of Tourism and Travel
Worms, Germany

Dr. Martin Buck
Messe Berlin GmbH
Competence Centre Travel and Logistics
Berlin, Germany

ISSN 1868-0127   e-ISSN 1868-0135
ISBN 978-3-642-27403-9 e-ISBN 978-3-642-27404-6
DOI 10.1007/978-3-642-27404-6
Springer Heidelberg New York Dordrecht London

Printed on acid-free paper

Springer is part of Springer Science+Business Media (www.springer.com)

Preface and Summary

Just as in previous years, talks and panel discussions were held at the 2011 ITB Berlin Convention on tackling issues dominating the global travel and tourism industry. This publication collates the convention's highlights in contributions from prominent industry professionals and academic specialists. Readers are likely to benefit from this comprehensive vision of the trends that will shape the global tourism industry's structure both today and in the future. This book is an indispensable source of information both for tourism and travel professionals as well as academics and students analysing the way in which global tourism and travel are set to go.

Chapter 1 centres on a variety of perspectives on the future of society, business and tourism. In his article, Ian Yeoman aims to illustrate what the future holds on the basis of the "ten certainties of change". Although you may not agree with him on every count, his conclusions are least enlightening and thought-provoking in the way he assumes tourism will develop in future decades. In the second article, Rohit Talwar draws on study findings in exploring how customer behaviour patterns might evolve, discusses the way technology could transform tomorrow's future travel experience and analyses the strategic implications facing the branded hotel sector. The World Travel Monitor represents the world's largest and most important travel survey; it conducts interviews in over 50 countries worldwide. Rolf Freitag and Dennis Pyka provide insight into worldwide and European travel behaviour and shed light on the trends that are currently dominating the market.

Chapter 2 covers selected aspects of destination management. Damba Gantemur gives insights into a niche destination with potential: Mongolia. This country captivates with unspoiled landscapes and authenticity – characteristics the hardened tourist is increasingly looking for. Klaus-Dieter Koch focuses attention on a well-established tourist destination – the Alps. He addresses the question of how tourism products and destination brand management can or must adapt to the challenges that face it both now and in the future. Burkhard Kieker sets out to explain Berlin's astronomic success as a tourist destination, provoking thought on what's been key in making city destinations the success they are. Andreas Reiter sums up studies carried out by ZTB Zukunftsbüro (a futures researcher) on the reasons for city destinations being such a popular option, concluding that cities need to be "worth living" in three ways: It must be possible to live them in the way locals do, create opportunities for interaction in public areas, and provide sustainability ecologically.

Chapter 3 looks at topical issues in marketing management. Ingo Markgraf, David Scheffer and Johanna Pulkenat apply the latest results of neuropsychological research in the tourism industry. The tourism industry is lagging behind other segments in its use of neuropsychological approaches. Using the Rewe Group as just one example, the article demonstrates how this new approach can be used in optimising the range of options tour operators can offer. Kirsi Hyvärinen, Dusanka Pavicevic and Dennis Hürten apply findings from change management to give tourist destinations a future-proof hallmark – something many destinations still need to do in face of global competition. They show how Montenegro has been so successful in the experience it has made in realigning, developing, marketing and selling itself as a product tourists wants.

Chapter 4 deals with social media and online bookings. David Perroud provides the latest findings from a study on an issue that's currently on the lips of all tour operators: Putting social media to their best advantage. A global, empirical study shows the part Facebook, Twitter and Tripadvisor play in travel decision-making and booking process: Which users employ social media and for what reasons? What changes can the travel industry expect? – Questions the empirical study can provide answers to. Using previously published panel data from GfK Travel, Alexandra Weigand gives insights into the way Europe's tourism market is going. She then goes on to analyse the differences that exist between booking behaviour at travel agencies and online. One of its innovative features will be the scope of preferences towards holiday hotels.

Chapter 5 concentrates on business trips and events. Stefan Fallert ventures a critical look at the efforts airlines are increasingly making to generate income from ancillary revenues, and demonstrates the impacts this will have on business-travel management. Oliver Graue pursues another dominant trend in the airline industry: The industry's increasing consolidation and its effects on business-travel management. For years, events have not only provided an increasing incentive for taking business trips but also for holiday travel; these too are taking on an increasingly important part in the basket of communication options companies have in attracting business. Markus Weidner defines the key factors that make events successful. Hans Rück discusses the aspect of event briefings. Presenting empirical results, he shows how briefings need structuring to ensure events are a success.

Chapter 6 takes a look at different aspects of the much-discussed aspect of corporate social responsibility. Taleb Rifai kicks off by giving reasons for the necessity of sustainability in the global tourism industry and shows best-practice examples in the public and private sector. Johannes Reißland and Petra Thomas discuss the significant Tourcert System, and analyse the effects CSR certification has on tourism companies. Dietrich Brockhagen devotes attention to a highly topical issue: The EU Emission Trading Scheme in the airline industry. He demonstrates how an airline can benefit from its own offset projects and voluntary passenger-based offsetting benefits – options airlines are still not likely to be aware of today. Using their own research results, Xavier Font and Andreas Walmsley finally illustrate which international hotel chains set a benchmark in the sense of satisfying corporate social responsibility.

Chapter 7, the last in this publication, concentrates on sustainable mobility. Roland Conrady provides an overview of the status quo in sustainable mobility, while illustrating future perspectives and challenges in relation to the various carriers. And it's not without good reason that Björn Dosch focuses on e-vehicles in the detailed look he takes on electromobility. The passenger car is the key form of transport for holidaymakers. Technical innovations won't succeed on the market if they aren't accepted by the man on the street. Herbert Lechner concludes by showing the results of a study carried out by GfK on user acceptance of electric vehicles in different European countries.

This work could not have been achieved without the remarkable dedication on behalf of the authors, who, for the most part, have taken on leading positions in the tourism industry. Special thanks go to Pia Viehl and Steven Gay from the Faculty of Tourism and Travel, Worms University of Applied Sciences. It is only through their tireless dedication, remarkable skill and well-founded expert knowledge that it has been possible to publish this work on time. Never losing sight of our high quality standards, this has made them instrumental in making this publication such a valuable source of information.

Frankfurt/Berlin, November 2011   *Roland Conrady*
                                    Worms University of Applied Sciences

                                                              Martin Buck
                                                          Messe Berlin

Contents

Future at a Glance

A Futurist's Perspective of Ten Certainties of Change **3**
Ian Yeoman

1 Introduction 3
2 A Futurist's Perspective of Ten Certainties of Change 3
3 Conclusion: Reflective Thoughts 17
References 18

**Hotels 2020 – Responding to Tomorrow's Customer and the
Evolution of Technology** **21**
Rohit Talwar

1 Hotels 2020: Beyond Segmentation 21
2 Drivers of Future Traveller Behaviour 22
3 New Customers, Emerging Destinations 22
4 The Quest for Unique and Individualized Experiences 23
5 Beyond Segmentation – the Rise of the Service Spectrum 23
6 Technology Enabled Customization 23
7 Personalizing the Hotel Experience 23
8 Evolving the Intelligent Hotel 24
9 Delivering a Total Service Model 24
10 Beyond Hotels – The Rise of Virtual Experiences 25
11 From Virtual to Immersive 26
12 Memory Transfer – The Ultimate in Immersive Reality? 26
13 Full Experience Transfer 27
14 Winning in Turbulent Times 27
15 Embracing New Management Thinking 30
16 Conclusion – Transforming Hotels from the Outside In 30

Forecasting Global and European Tourism 33

Rolf Freitag and Dennis Pyka

1   Introduction 33
2   Overview of World Tourism in 2010 33
3   European Tourism in 2010 36
4   Looking Forward 41

Destination Management – Culture, Landscape, Cities

Mongol Passion: History and Challenges – Can Tourism Be a Tool to Empower It? 49

Damba Gantemur

The 10 Brand:Trust Theses on the Future of Alpine Destination Management/Branding 57

Klaus-Dieter Koch

City Tourism – The New Magic of the Place 61

Burkhard Kieker

Liveable City – Sustainable Quality of Life as Success Driver for Urban Branding 69

Andreas Reiter

1   City Competition 69
2   Liveable Cities 71

Marketing and Distribution

The Needs of Package Tourists and Travel Agents – Neuromarketing in the Tourism Sector 79

Ingo Markgraf, David Scheffer, and Johanna Pulkenat

1   Introduction 79
2   Starting Position, Problems and Objectives 79
3   Solutions and Method 80

4   Study Design 82
5   Results 83
6   Discussing the Results and Practical Implications 91
References 91

Product Development, Marketing, and Distribution 93

Kirsi Hyvärinen, Dusanka Pavicevic, and Dennis Hürten

1   Introduction 93
2   Change Requires Leadership 95
3   Change Starts with Results 98
4   Case Study Montenegro 111
5   Conclusions 122
References 123

Social Media and Technology Tackle Tourism Industry

Social Media and Mobile Devices 129

David Perroud

1   Research Methodology and Sample 129
2   Usage of Social Media for Leisure Travel 130
3   Usage of Mobile Devices 132
4   Conclusions 133

Holiday Hotels and Online Booking Behavior – News from the GfK Research Panels 135

Alexandra Weigand

1   Methodology and General Background 135
2   Booking Trends in Europe 136
3   Booking Behavior Online vs. Offline with Focus on Hotels 137

Business Travel and Event Management

Ancillary Revenues in Air Transport – Gain and Pain 145

Stefan Fallert

1 My Very Personal Experience Lately 145
2 Ancillary Fees – A Great Alternative to Generate Revenues 145
3 Travel Manager's Main Challenge: Tracking 147
4 Customer Reaction 149
5 Solutions Under Way 149
6 What to Do Today? 150

**Concentrated Might in the Sky – Airline Alliances and
Travel Management** 151

Oliver Graue

1 Introduction 151
2 Current Airline Alliances Boom 152
3 Consequences for Travel Management and Business Travelers 153
4 Conclusion 155
References 156

Qnigge in Event Management 157

Markus Weidner

1 Introduction 157
2 Participants in the Auditorium 157
3 Defining Quality 158
4 The Auditorium's Expectations 159
5 Explaining the Term Qnigge® 160
6 The Consulting Philosophy of Qnigge® GmbH – Freude an Qualität 160
7 Introduction into Quality Management 161
8 Company Values Derive from Qnigge®-Values as an Example 162
9 How Well Do Employees Know Their Company's Values? 165
10 What Do You Know About the Term Quality? 166
11 Evaluating the Quality of an Event 167
12 Defining Chains of Service and Processes 170
13 Documenting Organizational Structure and Service Chains 171

14 Why Do Customers Change Their Service Provider? 172

15 Six Steps to Unmistakable Quality 172

16 Conclusion 173

References 173

Quality Events Need Quality Briefings – Professional Communications Briefing as a Key Factor for Creating Successful Events 175

Hans Rück

1 Introduction 175

2 Briefing: Term, Procedure, Meaning 179

3 Quality of Briefings in the Event Business: Results of an Empirical Study 183

4 Briefing as an Instrument for Event Planning 188

5 The Event Manager's Role: From Event Organizer to Communication
   Expert 194

6 Conclusion 195

References 196

Corporate Social Responsibility in the Tourism and Travel Industry

CSR and Sustainability in the Global Tourism Sector – Best Practice Initiatives from the Public and Private Sector 201

Taleb Rifai

1 Sustainability in the Tourism Sector 201

2 The Move Towards Sustainability 202

3 International Tourism in a Green Economy 202

4 Best Practices in the Public and Private Sector 203

5 Global Initiatives 204

6 Sustainability in the Tourism Sector: A Collective Responsibility 205

The CSR-System of Tourcert and the Effects of CSR Systems on the Company 207

Johannes Reißland and Petra Thomas

1 The Tourcert CSR Process 207

2 Effects on Travel Agents from Introducing the CSR System 211

References 214

Challenges Awaiting the Aviation Industry – Preparation for the Integration in the Emission Trading Scheme **215**

Dietrich Brockhagen

1   Introduction 215
2   Main Factors Causing CO_2 Emissions in Flight Operation 215
3   Optimizing the Climate Efficiency of Flights: Detailed City Pair
    Analyses for Airlines and Companies 217
4   Investment in Company-Owned Offset Projects and Voluntary
    Offsetting by Passengers: Multiple Benefits for Airlines 218
5   Conclusion 221
References 221

Corporate Social Reporting and Practices of International Hotel Groups **223**

Xavier Font and Andreas Walmsley

1   Taking Responsibility for Sustainability in the Hospitality Industry 223
2   Methodology 224
3   Results 226
4   Discussion 228
5   Conclusions 230
References 231

Eco-mobility

Status Quo and Future Prospects of Sustainable Mobility **237**

Roland Conrady

1   Introduction 237
2   Resource Consumption and Climate Impact of the Transportation
    Sector 237
3   Determining "Sustainable Mobility" 240
4   Digression: Basics of Power Generation 241
5   Concepts of Sustainable Mobility 243
6   Sustainable Mobility in Tourism Destinations 257
7   Conclusion and Outlook 258
References 259

Eco-mobility – Will New Choices of Electric Drive Vehicles Change the Way We Travel? 261

Björn Dosch

1   Introduction 261
2   Benefits of Eco- or Electric Mobility 262
3   Technical Solutions 264
4   Barriers for Electric Drive Vehicles 266
5   Perspectives of Electric Drive Vehicles 267
6   Conclusion 269
References 270

Customer Needs and Attitudes Regarding Electrical Cars 271

Herbert Lechner

1   Introduction 271
2   Potential Buyers 273
3   Reasons of Non-buyers 275
4   Conclusion 275

Authors

Brockhagen, Dr. Dietrich
  Executive Director
  atmosfair gGmbH
  Zossener Str. 55–58,
  10961 Berlin, Germany
  www.atmosfair.de

Buck, Dr. Martin
  Director
  Messe Berlin GmbH
  Competence Centre Travel & Logistics
  Messedamm 22
  14055 Berlin, Germany
  buck@messe-berlin.de

Conrady, Prof. Dr. Roland
  Worms University of Applied Sciences
  Department of Tourism and Travel
  Erenburger Str. 19
  67549 Worms, Germany
  conrady@fh-worms.de

Dosch, Björn
  Head of Traffic Department
  ADAC e.V.
  Am Westpark 8
  81373 Munich, Germany
  bjoern.dosch@adac.de

Fallert, Stefan
  Director
  Carlson Wagonlit Travel
  CWT Policy & Compliance Solutions Group Global
  31, Rue du Colonel Pierre Avia
  75904 Paris Cedex 15, France
  sfallert@carlsonwagonlit.com

Font, Dr. Xavier
   Director of Studies
   International Centre for Responsible Tourism
   Leeds Metropolitan University
   Bronte 113 Headingley Campus LS6 3QW
   Leeds, United Kingdom
   www.icrtourism.org
   www.xavierfont.info

Freitag, Rolf
   CEO
   IPK International
   Gottfried-Keller-Straße 20
   81245 Munich, Germany
   freitag@ipkinternational.com

Frohreich, Susanne
   Project Manager
   CODIPLAN GmbH
   Friedrich-Offermann-Straße 5
   51429 Bergisch Gladbach, Germany
   s.frohreich@codiplan.de

Gantemur, Damba
   Chairman
   STDC – Sustainable Tourism Development Center
   P.O.Box 166
   Ulaanbaatar 13, Mongolia
   gana@tourmongolia.com

Graue, Oliver
   Editor-in-chief BizTravel
   FVW Mediengruppe, BizTravel
   Wandsbeker Allee 1
   22041 Hamburg, Germany
   o.graue@biztravel.de

Hürten, Dr. Dennis
   Managing Director
   Trendscope – Market research and consulting
   Gottfried-Hagen-Straße 60
   51105 Cologne, Germany
   d.huerten@trendscope.com

Hyvärinen, Kirsi
  Independent Consultant
  Travel & Tourism Transformation Management
  c/o NTOCG
  Bulevar Svetog Petra Cetinjskog 130
  81000 Podgorica, Montenegro
  kirsi.hyvaerinen@gmail.com
  kirsi.hyvaerinen@montenegro.travel

Kieker, Burkhard
  CEO
  visitBerlin
  Am Karlsbad 11
  10785 Berlin, Germany
  burkhard.kieker@visitberlin.de

Koch, Klaus-Dieter
  Managing Partner
  Brand:Trust GmbH
  Findelgasse 10
  90402 Nuremberg, Germany
  kdk@brand-trust.de

Lechner, Herbert
  Division Manager GfK Mobility
  GfK Panel Services Germany
  Nordwestring 101
  90419 Nuremberg, Germany
  herbert.lechner@gfk.com

Markgraf, Prof. Dr. Ingo
  Til 2011, Dec. 31
  REWE Touristik GmbH
  Humboldtstr. 140
  51149 Cologne, Germany

  Since 2012, Jan. 1:
  Macromedia Hochschule für Medien und Kommunikation
  Richmodstr. 10
  50667 Cologne, Germany
  i.markgraf@macromedia.de

Pavicevic, Dusanka
  Senior adviser I
  Ministry of Sustainable Development and Tourism
  IV Proleterske brigade 19
  81000 Podgorica, Montenegro
  duska.pavicevic@t-com.me
  dusanka.pavicevic@mrt.gov.me

Perroud, David
  Partner, CEO
  m1nd-set
  Rue du Lac 47
  1800 Vevey, Switzerland
  info@ms-research.net

Pulkenat, Johanna
  180° visual systems
  Bleibtreustraße 51a
  10623 Berlin, Germany
  pulkenat@neuroips.com

Pyka, Dennis
  Head of World Travel Monitor
  IPK International
  Gottfried-Keller-Straße 20
  81245 Munich, Germany
  pyka@ipkinternational.com

Reißland, Johannes
  CEO
  forum anders reisen e.V.
  Der Verband für nachhaltigen Tourismus
  Wippertstr. 2
  79100 Freiburg, Germany
  johannes.reissland@forumandersreisen.de

Reiter, Andreas
  Director
  ZTB Zukunftsbüro
  Gilmsgasse 7
  1170 Vienna, Austria
  a.reiter@ztb-zukunft.com

Rifai, Dr. Taleb
World Tourism Organization (UNWTO)
Capitan Haya, 42
28020 Madrid, Spain
omt@unwto.org

Rück, Prof. Dr. Hans
Dean of the Department of Tourism and Travel
Worms University of Applied Sciences
Erenburger Str. 19
67549 Worms, Germany
rueck@fh-worms.de

Scheffer, Prof. Dr. David
180° visual systems
Bleibtreustraße 51a
10623 Berlin, Germany
scheffer@neuroips.com

Talwar, Rohit
CEO
Fast Future
19 Lyndale Avenue
NW2 2QB
London, United Kingdom
rohit@fastfuture.com

Thomas, Petra
Product Management & PR
a&e erlebnis:reisen – Begegnungen in Augenhöhe erleben!
Hans-Henny-Jahnn-Weg 19
22085 Hamburg, Germany
petra.thomas@ae-erlebnisreisen.de

Walmsley, Dr. Andreas
International Centre for Responsible Tourism
Leeds Metropolitan University
Headingley Campus
Leeds LS6 3QW, United Kingdom
a.walmsley@leedsmet.ac.uk

Weidner, Markus
  CEO
  Qnigge® GmbH – Freude an Qualität
  Am Hellenberg 15b
  61184 Karben, Germany
  mfw@qnigge.de
  Twitter: @qnigge

Weigand, Alexandra
  Senior Marketing Counsultant
  GfK Retail and Technology GmbH
  Nordwestring 101
  90419 Nuremberg, Germany
  alexandra.weigand@gfk.com

Yeoman, Dr. Ian
  Associate Professor of Tourism Futures
  Victoria University of Wellington, New Zealand
  Visiting Professor, European Tourism Futures Institute, Netherlands
  PO Box 600
  Wellington 6135, New Zealand
  ian.yeoman@vuw.ac.nz

Future at a Glance

A Futurist's Perspective of Ten Certainties of Change

Ian Yeoman

1 Introduction

In *Physics of the Future* Michio Kaku (2011) demonstrates that in 2100 we will control computers via tiny brain sensors and, like magicians, move objects around with the power of our minds. Artificial intelligence will be dispersed throughout the environment, and Internet-enabled contact lenses will allow us to access the world's information base or conjure up any image we desire in the blink of an eye. So what does this mean for tourism? Is the future a world of flying cars, teleportation and space ships? More realistically, what about peak oil and ageing populations? What about the middle classes of China and India, the debate about the climate change emerging technologies such as claytronics used in hotel design.

The purpose of this chapter is to illustrate the future through based upon ten certainties of change. The chapter will make you think, you might not agree with the illustrations but at least you will find them illuminating and thought provoking. The foundation of the chapter is ten certainties of change what will shape the destiny of tourism in the future decades. The ten certainties are drawn from the author's own experience and a number of research projects including *New Zealand 2050* (www.tourism2050.com), and *Tomorrows Tourist* (www.tomorrows-tourist.com).

2 A Futurist's Perspective of Ten Certainties of Change

From a knowledge perspective (Sparrow 1998), certainty is perfect knowledge that has total security. This is something that will happen, a continuity from the past or a high degree of precision. However, certain knowledge from a social sciences perspective is based upon experiences which the author believes is the truth. Therefore, certainty is more a matter of belief that change will occur based upon the patterns that "I" see. This is the role of the expert, or a futurist. Futurists understand the change, can see beyond the horizon. They have the ability to layer patterns of trends, draw conclusions in order to make predictions. This is the world of subjectivity in which the mind of the futurist is an interpretation device. Futurists deal with multiple types of knowledge and could be described as jack of all trades rather than experts in one particular field.

A futurist is a person who pieces together knowledge as a set of cognitive patterns which represents a pattern of the future and illustrates answers in picture frame of sense. Futurists help us make sense of the world. Futurists live in a world of emergent construction that changes as data emerges from the different tools, techniques and approaches to elicitation (Weinstein & Weinstein 1991). They often deploy triangulation of methodologies in order to capture and understand the world around them. Futurists never present objectivity but a range of alternatives of subjectivity. The research they are involved in presumes interpretation which Schwandt (1994) labels of constructivist interpretation (Schwandt 1994). This is an ontology that is predominantly local and specific in which the creation of knowledge is grounded in practise. This epistemology views knowledge in a subjective and transactional manner as merely suggesting directions along which to look, rather than providing descriptions of what to see (Blumer 1954). Therefore the following ten certainties are what "I" see as certain, it's as simple as that. Believe me or not.

2.1   The New Middle Classes

The world's economic balance of power is shifting rapidly, accelerated by the global recession in 2009. According to Stancil and Dadush (2010), China remains on a path to overtake the United States as the world's largest economic power within a generation and India will join both as a global leader by mid-century. Traditional Western powers will remain the wealthiest nations in terms of per capita income, but will be overtaken as the predominant world economies by much poorer countries. Prior to the Global Financial Crisis, the world's balance of economic power, as measured by real gross domestic product (GDP), was gradually shifting to the South and the East. Now, as industrialized countries slowly resume growth along their pre-crisis trajectory but do not fully recover output lost during the crisis, developing countries—whose output losses during the crisis were much lower—will accelerate out of the recession. In the coming years, the most successful developing countries, especially but not only those in Asia, will converge even more rapidly towards their advanced counterparts. So, what does this all mean for international tourism? For example, many observers (Sachs 2010) have pointed to future developments, such as the rise of China and India and other emerging economies as drivers of international tourism. With travel traditionally being strongly correlated to GDP and the ageing of the population in Western countries in the long term means falling GDP per capita, this will have a major impact on the outbound travel from Western countries such as Germany and France, which in the past has fuelled the growth of world tourism.

Table 1. Outbound departures from source country (,000)

Country	2000	2001	2002	2003	2004	2005	2006	2007	2008	2000-2008 outbound increase / decrease
Republic of Korea	5,508	6,084	7,123	7,086	8,826	10,080	11,610	13,325	11,996	118%
Argentina	4,953	4,762	3,008	3,088	3,904	3,894	3,892	4,167	4,611	-7%
Australia	3,498	3,443	3,461	3,388	4,369	4,756	4,941	5,462	5,808	69%
Brazil	3,228	2,674	2,338	3,225	3,701	4,667	4,625	4,823	4,936	53%
Canada	19,182	18,359	17,705	17,739	19,595	21,099	22,732	25,163	27,037	41%
China	10,473	12,133	16,602	20,222	28,853	31,026	34,524	40,954	45,844	338%
France	19,886	19,265	18,315	18,576	21,131	24,800	25,080	25,139	23,347	17%
Germany	74,400	76,400	73,300	74,600	72,300	77,400	71,200	70,400	73,000	-2%
India	4,416	4,564	4,940	5,351	-	7,185	8,340	9,783	10,868	146%
Indonesia	-	-	-	3,491	3,941	4,106	4,967	5,158	5,486	57%
Italy	21,993	22,421	25,126	26,817	23,349	24,796	25,697	27,734	28,284	29%
Japan	17,819	16,216	16,523	'13,296	16,831	17,404	17,535	17,295	15,987	-10%
Mexico	11,079	12,075	11,948	11,044	12,494	13,305	14,002	15,083	14,450	30%
Russia	18,371	18,030	20,428	20,572	24,507	28,416	29,107	34,285	36,538	99%
Saudi Arabia	-	-	7,896	4,104	3,811	4,403	2,000	4,126	4,087	-48%
South Africa	3,834	3,733	3,794	-	-	-	4,339	4,433	4,429	16%
Turkey	5,284	4,856	5,131	5,928	7,299	8,246	8,275	8,938	9,873	87%
UK	56,837	58,281	59,377	61,424	64,194	66,494	69,536	69,450	69,011	21%
USA	61,327	59,442	58,066	56,250	61,809	63,503	63,662	64,028	63,684	4%
Spain	4,100	4,139	3,871	4,094	5,121	10,464	10,678	11,276	11,229	174%
Belgium	7,932	6,570	6,773	7,268	8,783	9,327	7,852	8,371	8,887	12%

Source: UNWTO

Table one reflects the rise of outbound travel from the G20 countries, with the developing economies China and India leading the way. Between 2000–2008, outbound travel from China and India has grown 338% and 146% respectively. Other emerging economies such as Russia (99%), Indonesia (57%) and Turkey (87%), lead the way with percentage increases, whereas traditional developed economies of USA (4%), Japan (-10%) and Germany (-2%) show evidence of stagnation and decline. So, who will be the future tourist? China and India in 2050 will be the first and third largest economy in the world, with average GDP growth rates of 5.6% and 5.9% respectively per annum.

2.2   Where Have All the Germans Gone?

Post baby boomers in 2050, senior tourism could be a different proposition as many countries such as Germany, Italy, Spain and UK reform pension policy – therefore pensioners post 2050 will be economically less well off compared to previous generations and as a consequence the economic value of tourism will fall. In 2050, generation X & Y will probably retire with insufficient level of incomes. The average worker in 2050 as shown in OECD countries is expected to have a combined public – private pension benefits that represent less than 70% of final earnings. As a consequence, increased contributions with longer careers will become mainstream and retirement will become a fluid concept. Many private sector employers have already closed defined contribution schemes to new entrants and many public sector schemes have redefined many of the benefits.

Developed countries in the OECD have a large public sector with favourable pension provision compared to the private sector, as a consequence tourism has been one of the key benedictory of this policy. With falling birth rates and rising life expectancies, the commitment from government to pay for public sector pensions and benefits is declining. However, the future of pensions and public pensions depends only partially on demographics, it depends on the economic trends in employment and earnings that determine a national ability to pay for pensions in the future, and it depends on political factors that determine a country's willingness to pay. A study by Grimm *et al* (2009) for the German Federal Ministry of Economics and Technology, concluded the impact of demography on tourism was manageable and demand will be relatively stable. But a more long term perspective, ageing populations will become problematic for Germany and propensity to travel and actual travel pattern will fall due to less wealth per capita, health issues and stagnant house prices (Lohman & Danielsson 2004). A recent study by McKinney and Co (2009), examined Germany's demographic structure and observed that it is passing through an important point, its baby boomers are approaching retirement age and this will have an increasingly negative impact on wealth per capita in coming years.

This demographic pressure comes from two sources.

- Households available to create wealth will be limited by slowing population growth and reduced household formation.
- Financial asset accumulation will slow because the falling prime saver ratio will lower average savings per household and limit the pool of money that can be allocated to acquiring financial assets.

Adult population is increasing, with total population declining. While the total German population will begin to decline within the next two decades, the adult population (defined as people above 17) will still increase, with the group 55 and over growing the most at 1.3 percent per year. With an older population characterized by a higher household-to-population ratio than any other age group, the decline in the number of households will lag the decline in total population. Household formation is reduced. Lower rates of household formation will constrain aggregate wealth accumulation since there will be fewer households earning income and generating savings. Financial asset accumulation will be slowed by lower savings per household. Average savings per household will be reduced going forward because there will be fewer households in their prime saving years. The prime saving ratio measures the number of households in their peak savings years (defined as the 20-year age bracket with maximum household savings) relative to the number of elderly households (who save at lower rates or dissave) and therefore, captures the lifecycle effects caused by ageing. The ratio of German prime savers to elderly households has now passed an inflection point: the prime saver ratio will consistently decline over the next two decades, reaching 0.54 by 2024. This decline will impact the flow of savings from German households as older households save less.

Population ageing impacts wealth accumulation through lifecycle savings behaviour. Germany has a traditional "hump-shaped" lifecycle profile. The German household lifecycle savings curve is steeply inclined, reaches a peak in the late 40s, and then rapidly slopes down in the late fifties and retirement years. With income peaking at age 54 and the savings rate hitting the highest point earlier at 41, an average German household experiences peak savings around ages 45–49. This relatively early age for peak savings magnifies the impact of an ageing population since the decline in savings occurs at an earlier age than in other countries. Therefore, as the German population ages, it will experience a "lifecycle effect" on savings earlier than in other countries that have peak savings at later ages. While population ageing will affect German savings and GDP per capita accumulation in the next two decades, the impact may be appreciably larger after 2030 because of the impending sharp decline in population.

2.3   Tourist Identity, Behaviour and Attitudes

Rising incomes and wealth accumulation distributed in new ways alter the balance of power in tourism. The tourist is the power base which has shifted from the institution of the travel agent through the opaqueness of online booking for holidays and travel to the individual. At the same time, the age is rich for new forms of connection and association, allowing a liberated pursuit of personal identity which is fluid and much less restricted by influence of background or geography. The society of networks in turn, has facilitated and innovated a mass of options provided by communications channels leading to a paradox of choice. In the future market place, the tourist can holiday anywhere in the world whether it is Afghanistan or Las Vegas, to the extent the tourist can take a holiday at the North Pole or the South Pole and everywhere in between including a day trip into outer space with *Virgin Galactic* (Yeoman 2008). If 25m tourists took an international holiday in 1950, 903m took a holiday in 2008 (Yeoman 2008). Why? The growth in world tourism is founded on increases in real household income per head, which doubles every 25 years in OECD countries. This increase in disposal income, allows real change in social order, living standards and the desire for a quality of life with tourism at the heart of that change. Effectively, consumers want improvement year on year, as if it was a wholly natural process like ageing. That change in disposable income, has meant greater and enhanced choice for tourists.

That tourist has demanded better experiences, faster service, multiple choice, social responsibility and greater satisfaction. Against this background, as the world has moved to an experience economy in which endless choice through competition and accessibility because of the low cost carrier, what has emerged is the concept of *fluid identity*. This trend is about the concept of self which is fluid and malleable in which self can not be defined by boundaries, in which choice and the desire for self and new experiences drives tourist consumption. The symbol of this identity, is the fact that the consumer on average changes their hairstyle every 18 months according to research by the Future Foundation (2007), from a tourist perspective it is about collecting countries, trying new things and the desire for constant change. It means the tourist is both comfortable with a hedonistic short break in Las Vegas or a six month ecotourism adventure crossing Africa. This fluid identity makes it difficult for destinations to segment tourists by behaviour or attitude as it is constant and fluid. However, as wealth decreases that identity becomes more *simple,* a new thriftiness and desire for simplicity emerges (Flatters & Wilmott 2009). This desire for simplicity is driven by inflationary pressures and falling levels of disposable incomes, squeezing the middle class consumer. As the economies of wealth slows down, whatever the reason, new patterns of tourism consumption emerge, whether it is the desire for domestic rather than international travel or what some call the *stayvacation.* A fluid identity means tourists can afford enriching new experiences and indulge themselves at premium 5 star resorts. They can afford to pay extra for socially conscious consumption, whereas a simple identity means these trends have slowed, halted or reserved. As resources become more scarce, a mindset of a whole generation of tourists changes behaviour. Between now and 2050 the world will go through a cycle of economic prosperity and decline which is the nature of the economic order. When wealth is great, a fluid identity is the naked scenario, however, when a recession emerges, belts are tightened, tourists like other consumers search for a simple identity.

2.4   How Technology Will Change Everything

The internet is one of the main drivers of product design as many mobile devices are increasingly equipped with mobile internet capabilities. Connectivity to the internet allows faster and more immediate access to information. A survey conducted by TNS Global (2008) indicated that many see the internet as "an encyclopedia of information", where 3 out of the top 5 activities engaged by online users are related to information gathering. Survey results also indicate that 81% of respondents used a search engine to find information, 63% researched a product or service before, 61% visited a brand or product's website and 50% used a price comparison chart. These figures suggest that consumers are increasingly turning towards the internet to obtain information on products, brands and pricing. Within the tourism industry, internet is being targeted to become the most important channel for holiday sales, information and recommendation where 2 out of 5 reservations are completed online and 55% of all European travellers use the internet for information about their travel destination, travel providers and special offers (Isabel, 2009). Recognizing this trend, Tourism New Zealand in 2007 shifted its marketing activities from predominantly print media to embrace digital and screen technology. This includes advertising through television, cinemas, outdoor screens and billboards and more significantly, the internet and social media (Tourism New Zealand, 2010).

Technology has become part of our everyday lives, creating a digital society. While one of the main reasons for this is due to the exponential advancement in technology, another key driver is due to the presence of the digital generation (Generation Y onwards), and their demand for fast, innovative technology products. High-speed broadband with larger bandwidth have allowed greater capacity of network traffic and data sharing while new gadgets increasingly equipped with mobile internet reflects the level of demand and comfort societies have towards technologies. This trend is further echoed when Amazon's sales of books for its e-reader, the Kindle, outnumbered sales of hard covered books (Miller, 2010). Technology has also allowed the development of online user-generated content, altering the way information is provided, gathered and perceived. Information provision has evolved from the traditional single-directional push of information from suppliers to consumers to a multi-directional share of information between suppliers and consumers, and between consumers themselves. Deloitte predicts that in 2011, more than 50% of computing devices sold globally will not be PCs. Instead, sales of smartphones and tablet computers would come to 425 million, well above the sales of 390 million PCs (Yeoman 2012). This implies that user-generated content will increasingly penetrate the online world of information, reflecting two future scenarios of a "Free Information Society" and "Real Information Society" proposed by Yeoman & McMahon-Beattie (2006). A free information society highlights that information is freely available and consumers no longer need to purchase information, whereas a real information society reflects how technology supports personal information rather than replacing it.

In today's society, digitalised information is the norm. Many guidebooks such as Lonely Planet have embraced mobile devices by providing digitalised guidebooks through the format of mobile applications designed for smartphone operators like Nokia, Apple, Google and Android (Lonely Planet, 2011). However, the continuous development of technology is bringing societies into a flip point, where technologies increasingly become more integrated in our daily routine. Driving this is ubiquitous computing – a concept opposing virtual reality. Ubiquitous computing refers to technologies which interact with humanity out in the open rather than users connecting through the computer; it is the interaction of one user with many interfaces through technology that is interwoven into the external environment. This concept puts forth many possibilities of interaction with information technologies without the use of devices, for example the possibility of gathering information of a subject of sight through a pair of ubiquitous contact lens. As technology slowly recedes into the background and becomes an invisible interaction in our daily lives, the future of information provision may no longer require the need of mobile devices.

2.5   The Complexity of Science

There appears to be an exponential growth involving the usage of the term "complexity" in the scientific literature. As society develops and grows, it finds new solutions, discovery is the heart of medicine, technological revolution seems endless and therefore mankind seems to be facing a world in which the pace of discovery is infinite. As a result, complexity and the pace of discovery is changing the world of science, technology and medicine, to the extent that simple human mortals cannot keep pace with this change, as a consequence the meeting's industry has been a beneficiary. For example, medical doctors instead of meeting every five years to keep abreast of change have to meet every two years instead (HCEA 2009). Why? Medical treatments for cancer have radically improved survival rates after five years from 50% in 1975 to 66% in 2002. In other fields of medicines, new and better treatments are emerging. Over the last decade we have seen life expectancy massively increase for patients diagnosed with HIV due to new treatments and drug therapy combinations. Shoemaker and Shoemaker (2009), notes that 1 in 5 people in advanced economies by 2030 will probably celebrate their 100th birthday in their lifetime. The authors go on to say that the average 50-year woman living in the USA in 1990 could look forward to an average of 31 additional years of life. If we assume for a cure for cancer, this increment beyond 50 years of age grows to 34 years; adding a cure for heart disease amounts to 39 years. After we conquer strokes and diabetes, the increment rises to 47, yielding a full life expectancy of 97 years of age. No one knows for sure how the boundary of death can be pushed, but optimistic scientists consider 130 years to be feasible by 2050. Basically, the more complexity we have the world, the more we have to discuss it, understand it and the meetings industry is the beneficiary.

2.6   The End of Human Trafficking

David Levy (2007), suggests in his book *Love+Sex with Robots* that by 2050 technological advancement will allow humans to have sex with androids, something akin to the *Stepford Wife* concept of a woman with a perfect body and who can perform great sex with a man. In 2006 (Levy 2007), Henrik Christensen chairman of EURON, the European Robotics Research Network predicts that people will be having sex with robots in five years and in 2010 the world's first sex doll was showcased at the AVN Adult Entertainment Expo in Las Vegas. Priced between $7000 and $9000 US Roxxxy is a truly interactive sex doll offering a range of replicated personalities from frigid Farah to Wild Wendy.

Robot sex offers a solution to a host of problems associated with the sex trade. Given the rise of incurable Sexual Transmitted Infections (STIs) including emergent strains of gonorrhea and HIV/AIDS throughout the world and the problem associated with human trafficking and sex tourism, it is likely that we will see an increase in demand for alternative forms of sexual expression. In 2050, Amsterdam's red light district will all be about android prostitutes who are clean of sexual transmitted diseases, not smuggled in from Eastern Europe and forced into slavery. The city council will have direct control over android sex workers controlling prices, hours of operations and sexual services. Android prostitutes will be both aesthetically pleasing and able to provide guaranteed performance and stimulation for both men and women.

2.7   The Future of Cities

By 2050, most regions of the world will be predominantly urban as 193,107 new city dwellers are added to the world's urban population every day. This translates to slightly more than two people every two seconds (UN-HABITAT 2008), reflecting that the world's urban population will swell to almost 5 billion in 2030 and 6.4 by 2050. Cities have long been the centre of tourist activity, from the early times of civilisation through to their highly developed state in the global economy. Cities hold a particular fascination for tourists, from the vast highly developed metropolitan cities like Los Angeles to small historic cities like Durham in the North East of England. Professor Stephen Page of Stirling University argues (Page & Connell 2006), that urban tourism is arguably one of the most highly developed forms of tourism at a global scale, by a post industrial society and an affluent society. According to Yeoman (2008), the growth of world tourism in the last decade has been due to inter-regional travel rather than inter-continental travel fuelled by inter-city short breaks and the budget airlines. Cities have become activity places for culture, sports and amusements as well as offering leisure settings with physical characteristics and socio cultural features. Many planners have turned to tourism as a means for urban regeneration. For example, the redevelopment of the London Docklands in 1980s onwards has been marketed as an example of London's vibrant tourism economy. The urbanisation of tourism in Los Angeles has become a key component of the city's economy and integral part of the city's urban development. In recent years, I-Max screens, themed environments, mega stores, theatres, museums and sports venues have displaced marble-clad office towers in Los Angeles as it has become a sprawling metropolis of entertainment in the era of the experience economy. However, what is the future for the city of Los Angeles? According to studies by Scott et al (2004), California and the cities of San Diego and Los Angeles could be classified as an optimal climate for tourism and all year round visitation. But what does the future hold given warmer climates, rising sea levels, water shortages, peak oil and the continuing trend of urbanization? Scott's et al (2004), study examined climate change scenarios for tourism in U.S cities through 2030 to 2080 and found that Los Angele's tourism would be marginally better off in the winter months but overall would move from "excellent" to "marginal/unfavourable", as the climate would become unbearable for tourists. If so, how will the city of Los Angeles adapt and mitigate for such change given the certainty of climate change? Would Los Angeles in 2050 be something akin to *Logan's Run* (Nolan & Johnson 1967), as portrayed in the classic science fiction film, where life is controlled and managed within a domed complex?

This chapter takes a futuristic perspective on what tourism in urban Los Angeles will be and the relationship to California's hinterland in 2050.

More than 70% of the population in developed worlds are living in an urban environment. According to the United Nations, this urban population is expected to remain largely unchanged in the next two decades, increasing from nearly 900 million people in 2005 to nearly 1.1 billion by 2050 – growth resulting from external in-migration rather than natural population growth. North American cities grew the fastest among all cities in the developed world between 1990 and 2000, particularly cities in the United States, which grew an average of 1% per annum. Las Vegas – the gambling and tourist resort in the state of Nevada – grew at the annual rate of 6.2%, and the city of Plano on the outskirts of Dallas, Texas, saw growth rates of 5.5% per year due to migration from other parts of the United States. As the US's most populous state, California's population increased from 30 million in 1990 to 336.5 million in 2004, growing at 600,000 people per year. According to the California Department of Finance (Yeoman 2008), the state's population is projected to exceed 48 million by 2030 and reach 60 million by 2050. These projections indicate that the majority of Californians will continue to reside in Southern California and Los Angeles will remain the most populous county in California. Los Angeles often abbreviated as L.A. and nicknamed *The City of Angels*, has an estimated population of 3.8 million, its metropolitan area with 12.9 million residents, and spans over 498.3 square miles (1,290.6 km^2) in Southern California.

2.8   The Price of Food

According to Evans (2008), the main drivers for food inflation are: the rising costs of agricultural inputs and energy, water scarcity, decreased land availability and climate change.

Today's global agricultural system is predicated on the availability of cheap, readily available energy, for use in every part of the value chain: both directly (e.g. cultivation, processing, refrigeration, shipping, distribution) and indirectly (e.g. manufacture of fertilizers, pesticides). World oil prices peaked in 2008 and will remain relatively high in the long term as the world has passed the point of peak oil. In addition, since food can now be converted into fuel, there is effectively an arbitrage relationship between the two, implying an ongoing linkage between food and fuel prices. Secondly, water scarcity is likely to become a more pressing issue. Global demand for water has tripled in the last 50 years; 500 million people live in countries chronically short of water, a number likely to rise to 4 billion by 2050.

Thirdly, is the issue of land availability? Some commoditys analysts argue that although historical increases in demand have been met through increasing yields, the future would require an expansion of acreage. However, this will be expensive, given the infrastructure investment involved; there may also be diminishing returns, since much of the best land is already under cultivation. Above all, there is simply increasing competition for available land, including food, feed, fibre (e.g. timber, paper), fuel, forest conservation, carbon sequestration and urbanization, on top of high rates of soil loss to erosion and desertification. The Food and Agricultural Organisation (FAO) of the United Nations estimates that there is at most 12% more land available that is not already forested or subject to erosion or desertification, and that 16% of arable land is already degraded.

The fourth and perhaps most fundamental factor is climate change. The International Panel on Climate Change (Parry 2007) projects that global food production could rise if local average temperatures increase by between 1 and 3 degrees Celsius, but could decrease above this range. This projection is before extreme weather events are taken into account; and the IPCC judges that extreme weather, rather than temperature, is likely to make the biggest difference to food security. It estimates that glacial melting will affect agriculture and many Himalayan glaciers could disappear by 2035, bringing catastrophic outcomes for Chinese and Indian agriculture industries. It assesses that "climate change increases the number of people at risk of hunger", and will lead to an increase of the number of undernourished people to between 40 million and 170 million.

As a consequence, food tourism becomes the new luxury as availability and scarcity drive demand for this new luxury product. Food pervades our lives from almost any perspective we care to consider; it is a primary feature of everyday life—we must find, purchase or prepare food and eat every day to stay healthy and alive; food permeates our relationships—we eat with others, and in particular and symbolic ways; food infiltrates our language—the images and metaphors of food surround us (I'm fed up with you, you make me sick, etc). Food reflects our position and status—whether we eat minced mutton, rabbit ragout or pasta primavera; food pervades popular culture—evidenced by the large number of cooking programmes on television, the ubiquitous cooking recipes and restaurant reviews. Food is a critical contributor to a human's physical well-being, a major source of pleasure, worry and stress, and the "single greatest category of expenditure".

2.9   The Modernity of Sustainable Hotels

The sustainability agenda is becoming the increasingly important priority for countries around the world, as Matteo Theo says (Putz-Willems 2009: 67)

> *The term sustainability could almost be defined as the buzz word of the early 21ˢᵗ century. Sustainability is the talk of the town, not only in the economy or in politics, but also in the construction industry. Considering everything closely however, only one aspect of sustainability catches the industry's attention, the ecological perspective. In this respect, the use of different resources throughout a building's complete lifecycle is balanced. This ecological balance corresponds to the materials used from production to demolition and even the essential resource needs for building management throughout the whole usage*

period. The two other aspects of sustainability are often forgotten, the economic and socio-cultural points. The building should maximise its potential to reduce maintenance costs, if possible even generate profits and should lose very little in value. On the other hand, it should also cater to the user's wellbeing in regard to health and comfort aspects as well as be aesthetically pleasing.

Architecture is important for offsetting the negative aspects of hotel buildings by enhancing efficiency of resources as well as making the design aesthetically pleasing. The modernity of architecture finds a home in futurism, in which sustainable design provides an opportunity for pushing out barriers and thinking beyond the present. Futurism and technology sit together, and given the future Chinese tourists will have an expression for newness captured by Generations Y and Z. Innovation in architecture has a linimal space where reality and science fiction are blurred, new ideas and concepts are emerging that shifts the concept of a hotel bedroom. What we think of science fiction, in fact comes true. Whether it is self cleaning devices, mood zones, claytronics and gestural interfaces – all concepts more akin to sci-fi film *Minority Report* than the hotel in 2050. Technology is being largely infused within new modern hotels for two main reasons: to improve the efficiency of hotel operations and to cater better to the evolving new segments of hotel users. The future hotel will become a technologised space, shifting from its original labour-intensive nature. This trend is driven by new innovations such as nanotechnologies in self-cleaning devices, robot room attendants, hi-technology wall-mounted toilet designs and elements of lighting, ambience and furniture that allow guests to recreate their personal space to suit their moods.

As a report by Amadeus (Talwar 2010) points out, at the core of any hotel stay, guests will want to exercise most choice when it comes to the location and contents of their room. The range of options would need to include the floor, corridor positioning, view, room dimensions, shape, number of windows, size of bathroom, and the type, amount and layout of furniture. By 2020 (Talwar 2010), modular, intelligent furniture with built-in memory will remember a guest's preferred settings and adapt to changes in body posture. Taking this concept one stage further, claytronics will allow furniture to re-configure themselves based upon programmable matter.

At the heart of a hotel room, customers want to choose from a range of different beds, pillows, linens and amenities at different quality levels and price points. Some require transparency on the environmental footprint of the supply chain of everything that goes into their room. Guests want the ability to control environmental factors such as temperature, lighting and even the colour of the walls. Choice could also be extended to the type of artwork displayed on the walls or for the provision of digital photo frames to display the guest's own choices. As technology advances and intelligent wallpapers emerge, so guests may be able to configure the room décor on arrival or download their preferred design beforehand.

The Citizen M Hotels (www.citizenm.com) in Amsterdam combine several innovations in room technologies to provide the guest with a chic and due to the small
size, affordable experience. The pod like size requires an innovative approach to
space management – for example there isn't room to move around the bed to
change the sheets. Citizen M has applied to patent a system whereby the whole
mattress can be pulled up to the front of the bed vertically. The used sheets fall off
and the clean sheets can be hung up on the two upper corners.

The rate of advance in technology and the likely emergence of high bandwidth
mobile devices means guests may want a room with no technology (just to get
away from everything). Others may simply be looking for a display screen or
surface to project a larger image from their own device. Those who do, the hotel
to supply the technology may wish to specify the channels they would like to view
and request a holographic TV. A guarantee of the chance to try out the latest
gadgets may become a brand differentiator and attract a particular type of customer. Some guests may want to book the opportunity to test out a new product or
schedule a session with a technology advisor to help them master what they already have. Given the trend towards individualism and life with technology,
Trump Soho in Manhattan (www.trumpsohohotel.com) boasts as an exemplar
of this trend. Central to its guestrooms and suites is the energy saving "Control4
Suite System" which enables guests to control ambient temperature, lighting,
curtain drapes and entertainment options with a remote device. Guests can set
their own room preferences using the green feature button. This offering is augmented by flat screen televisions, a home iPhone/iPod and docking station as well
as optional in-room computers and personalized stationery. The offering is completed with a Nespresso coffee maker in each guestroom and suite. Moving into
the future, technology will play and even more important part in the hotel bedroom, the use of gestural interfaces will change room control panels. 3D hologram
TVs will become the norm. The application of technologies is probably unimagin-
able and occurring very fast. One example is the medical mirrors designed by MIT
researcher Ming-Zher Poh (Chandler 2010), which will advise consumers of
health requirements, how they feel and what they could order off the room service
menu courses of actions. The system works by:

> ... measuring slight variations in brightness produced by the flow of
> blood through blood vessels in the face. Public-domain software is
> used to identify the position of the face in the image, and then the
> digital information from this area is broken down into the separate
> red, green and blue portions of the video image. In tests, the pulse
> data derived from this setup were compared with the pulse deter-
> mined by a commercially available FDA-approved blood-volume
> pulse sensor.
>
> Chandler (2010)

2.10 Peak Oil

Peak oil is the point in time when the maximum rate of global petroleum extraction is reached, after which the rate of production enters terminal decline. This concept is based on the observed production rates of individual oil wells, and the combined production rate of a field of related oil wells. The aggregate production rate from an oil field over time usually grows exponentially until the rate peaks and then declines—sometimes rapidly—until the field is depleted. According to Becken (2010), the world is now in the period of peak as majority of the studies about Peak Oil suggest a point between now and 2022 (Figure 1).

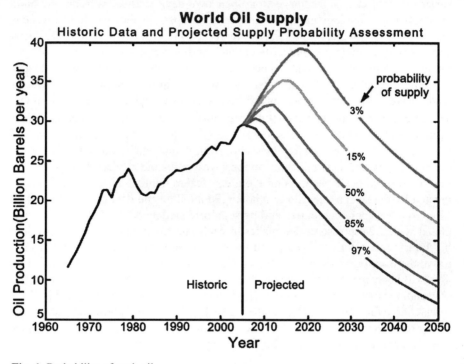

Fig. 1. Probability of peak oil

Source: Becken 2010

A point of Peak Oil signifies a period of oil depletion and exponential decline with the end of oil predicted to be pessimistically 2054 and optimistically 2094.

The US Department of Energy (DoE) calls oil "the lifeblood of modern civilisation" (Hirsch etc al 2005). Around 86 million barrels (13.7 billion litres) are consumed each day. Oil supplies 37% of the world's energy demand and powers nearly all of the world's transportation. Without it, production and trade would grind to a halt. Studies have shown that GDP growth is very strongly related to increased use of oil (Smith 2010). When the price of oil increases, the cost of nearly all economic activity rises. This often induces recessions. High oil prices have been associated with three major periods of economic recession in the past 40 years, including the lead-up to the recent global economic crisis. The world's oil production capacity may not be sufficient to match growing demand in coming years. When that happens, a price spike may be triggered, with major detrimental effects on economies, especially economies dependent on tourism (Yeoman et al 2007).

3   Conclusion: Reflective Thoughts

Some would say that the only certainties in life is birth and death, everrything else that happens in between is uncertain. Uncertainty stems from risk, a lack of understanding or lack of familiarity. All that "I" am saying is, that the ten certainties presented in this chapter are what "I" truthfully believe is going to happen, given that change is inevitable. For the tourism industry it is about dealing with that change as the only real certainty is that "you" can only live in the future as the past has already happened. The purpose of this chapter is about making you think about the future, "I" hope the words have made this possible, if not try these questions:

- How will an ageless society impact upon you business or markets?

- How are the behaviours and attitudes of Indian and Chinese tourists different from your present market? How do you prepare for such change?

- What do you think the tourist will be doing more of or less of in the future?

- How does technology change your distribution channel?

- Would tourists really have sex with a robot?

- Does the greater urbanization of society result in tourists wanting a rural authentic experience?

- How do you sell in a market size of one?

- What if the oil ran out?

- Reflecting on everything written in this chapter, what are the most important points to you?

- What are your ten certainties for the future?

- What if, the future tourist had less monies in the future? How will it change your market or business?

- How do you position food as a luxury experience?

And finally, just enjoy the future as much as "I" do.

References

Becken S (2010) The Future of Oil. Victoria Management School Seminar Series, Wellington. 28th July

Blumer H (1954) What is Wrong with Social Theory? *American Sociological Review*. Vol 19, pp 3–10

Chandler T (2010) Your Vital Signs on Camera.Accessed on the 27th Febraury 2011 at http://web.mit.edu/newsoffice/2010/pulse-camera-1004.html

Evans A (2008) Rising Food Prices. Chatham House, London. Accessed on the 5th November 2010 at www.chatham.org.uk

Flatters P & Wilmott M (2009) Understanding the Post Recession Consumer. *Harvard Business Review*. July-August, pp 106–112

Future Foundation (2007) NVision Central Scenario UK. Future Foundation, London

Grimm B, Lohmann M, Heinsohn K, Richter C & Metzler D (2009) The impact of demographic change on tourism and conclusions for tourism policy. A study commissioned by the Federal Ministry of Economics and Technology. Accessed at http://www.bmwi.de/BMWi/Redaktion/PDF/Publikationen/Studien/auswirkungen-demographischer-wandel-tourismus-kurzfassung-englisch,property=pdf,bereich=bmwi,sprache=de,rwb=true.pdf on 1st July 2010

HCEA (2009) Top Healthcare Meetings and Destinations. Accessed on 10th January 2010 at http://www.hcea.org/

Hirsch R, Bezdeck R, & Wendling R (2005) Peaking of World Oil Production: Impacts, Mitigation & Risk Management. Accessed on the 5th December at http://www.netl.doe.gov/publications/others/pdf/oil_peaking_netl.pdf

Isabel B (2009) Online Travel: The Internet is the key sales channel for tour operators. *ITB Berlin Special Press Release,* March 1–9

Kaku M (2011) Physics of the Future: How Science Will Shape Human Destiny and Our Daily Lives by the Year 2100. Allen Lane, London

Levy D (2007) Love + Sex with Robots. Harper Perennail, New York

Lohman M & Danielsson J (2004) How to Get the Future of Tourism Out of Today's Consumer Survey. Prospects for Senior and Kids Travel in Germany. 7th International Forum on Tourism Statistics, Stockholm, Sweden. 9–11th June. Accessed on 10th October at http://www.tourismforum.scb.se/Consumer_Surveys.asp

Lonely Planet (2011) *Mobile*. Accessed at http://www.lonelyplanet.com/mobile/ on 16th February 2011

McKinsey & Co (2009) Germany: Storm Clouds Gathering. Accessed on the 30th October 2009 at http://www.mckinsey.com/mgi/publications/demographics/germany.asp

Miller CC (2010) E-Books Top Hardcovers at Amazon. Accessed at www.amazon.com on the 19th July 2010

Nolan W & Johnson G (1967) Logan's Run. Corgi Books. New York

Page S & Connell J (2006) Tourism: A Modern Synthesis. Thomson; London

Parry M, Canziani J, Lindem P & Hanson C (2007) Contribution of Working Group II to the Fourth Assessment Report of the Intergovernmental Panel on Climate Change, 2007. Cambridge University Press, Cambridge

Putz-Willems M (2009) Hotel Architecture: Less Is More – Or: Natural Sustainable. In Conrady, R & Buck, M (2008) Trends and Issues in Global Tourism. Springerlink, Berlin

Sachs A (2010) The Travel Gold Rush 2020. Accessed on the 1st January 2010 at www.oef.com/free/pdfs/travelgoldrush2020.pdf

Schwandt TA (1994) "Constructivist, Interpretivist Approaches to Human Inquiry". In Denzin N & Lincoln Y (eds) (1994) Handbook of Qualitative Research, pp 118–137. Sage, London

Scott D, McBoyle G & Schwartzentruber M (2004) Climate Change and the Distribution of Resources for Tourism in North America. *Climate Research.* Vol 27 No 2, pp 105–117

Shoemaker P & Shoemaker J (2009) Chips, Clones and Living Beyond 100. Pearson Education, London

Smith C (2010) The Next Oil Shock. Parliamentary Library. Accessed on the 6th December at http://www.parliament.nz/en-NZ/ParlSupport/ResearchPapers/4/6/a/00PLEco10041-The-next-oil-shock.htm

Sparrow J (1998) Knowledge in Organisations. Sage, London

Stancil B & Dadush U (2010) The World Order in 2050. Accessed on the 30th December 2010 at http://carnegieendowment.org/files/World_Order_in_2050.pdf

Talwar R (2010) Hotels 2020: Beyond Segmentation. Accessed at http://www.amadeus.com/hotelit/beyond-segmentation.html on the 20th March 2011

TNS Global (2008) Digital Life Digital World. Accessed at http://www.tnsglobal.com/_assets/files/TNS_Market_Research_Digital_World_Digital_Life.pdf on 15th February 2011

Tourism New Zealand (2010) *100% Web Ready.* Accessed at http://www.tourismnewzealand.com/campaigns/consumer-marketing/100percent-web-ready-/ on 18th February 2011

UNHABITAT (2008) State of the World's Cities 2008/2009. Earthscan, London

Weinstein D & Weinstein MA (1991) George Simmel: Sociological Flaneur Bricoleur. *Theory, Culture & Society,* Vol 8, pp 151–168

Yeoman I & McMahon-Beattie U (2006) Tomorrow's tourist and the information society. *Journal of Vacation Marketing, Vol* 12 No 3, pp 269–291

Yeoman I, Lennon J, Blake A, Galt M, Greenwood C & McMahon-Beattie U (2007) Oil Depletion: What Does This Mean for Scottish Tourism. *Tourism Management.* Vol 25, No 5, pp 1354–1385

Yeoman I (2008) Tomorrows Tourist. Elsevier, London

Yeoman I (2012) 2050: Tomorrows Tourist. Channelview, Bristol

Hotels 2020 – Responding to Tomorrow's Customer and the Evolution of Technology

Rohit Talwar

Rohit Talwar is a futurist for the global travel industry and author of the Hotels 2020 study commissioned by Amadeus. In this article, Rohit draws on the study findings to explore how customer behaviours might evolve, discusses how technology could transform the future travel experience and explores the strategic implications for the branded hotel sector.

1 Hotels 2020: Beyond Segmentation

The future of travel and tourism is being driven by a complex set of converging forces which are forcing the industry to think about how they might reshape the experience. As hoteliers respond to increasing economic turbulence and start developing strategies for the next decade, it is critically important to scan ahead, understand these key external drivers of change and identify the emerging opportunities that could shape the strategic agenda for the sector. The **Hotels 2020: Beyond Segmentation** study set out to explore these "future factors" and explore the implications for hotel strategy, brand portfolio, business models, customer targeting and innovation. The study placed particular emphasis on the impact of increasing personalization, emerging technologies and a changing economic outlook.

We used a range of primary and secondary research techniques to identify key drivers of change, emerging opportunities and evolving guest and traveller expectations. We also scanned widely for examples of how the hotel sector in particular was already innovating, for ideas on what future innovations might look like and for technology developments that could impact the industry. We used these to suggest a range of future scenarios covering how guest behaviour might change, what the implications might be for the services, facilities and technologies hotels would require and how hotel groups might evolve their strategies, business models and market positioning. We then tested guest and industry responses to these scenarios in a global survey with just over 600 respondents.

2   Drivers of Future Traveller Behaviour

The role of travel in our lives is changing. In a survey conducted as part of the study, 83% of respondents strongly agreed or agreed that *"People will view travel as a right rather than a luxury and consider it an increasingly important part of their lives"*. As individuals, we are facing increasing pressure on our available leisure time, which is driving the desire for unique and personalized experiences. We are also not completely logical and consistent in our behaviours. On the one hand we want constant connectivity, yet on the other we are looking for stories to tell our friends about the remote new "unspoilt" destinations we've discovered!

Concerns over environmental impacts of tourism, and greenhouse gas emissions from transport in particular, are encouraging travellers to think more carefully about their ecological footprint. In the study 82% agreed that *"Environmental considerations will play an increasing role in the choice of business and leisure hotels"*. For tourism destinations, environmental factors are forcing a rethink of how many tourists to allow in, what to charge and who to target. Political upheavals and security concerns are also leading some travellers to rethink the desirability of visiting certain locations. Finally, innovation is reshaping the travel experience by – for example – enabling shorter flying times.

3   New Customers, Emerging Destinations

The study found that declining real incomes in the developed world coupled with rising affluence in the emerging nations are reshaping the visitor profile and driving a shift in emphasis on which outbound markets to target. In the survey, 75% believed that *"The Asian middle classes will make up the largest share of international travel"*. There is a growing understanding of the value and cost of servicing different groups of customers. While some leave a room cleaner than when they arrive, others can be expensive to service. Hence, 97% felt that *"Hotels will increasingly consider factors such as cost of servicing, level of spend and average length of stay when targeting potential customers in different geographic markets"*.

As a consequence of the expected growth of developing destinations, 79% agreed that "Heavy investment in emerging tourism markets will widen traveller choice, increase competition and potentially drive down prices and profit margins across the spectrum of hotels". The rise of new competitors was also expected to see new strategies for destination marketing emerge, with 82% expecting that by 2020 "City or country based alliances are likely, resulting in preferential marketing/pricing of certain destinations". Despite its growing economic power, only 32% felt China would become the world's top tourist destination by 2020.

4   The Quest for Unique and Individualized Experiences

The core finding of the study is that our guests will want a more personal, connected and informed experience. Customers are getting used to increasing choice and personalization – from the way in which we communicate to them to the choice of seat in the plane – they want it "their way". The danger in an increasingly fragmented world is to still try and think of customers as segmented groups as we have in the past. However, an increasingly diverse customer base and technology enabled personalization mean that guests no longer fit into the clean segments of yesteryear. Survey respondents agreed and 71% believe that *"Traveler motivations will become increasingly fragmented and diverse and harder to segment into clearly definable customer groupings"*.

5   Beyond Segmentation – the Rise of the Service Spectrum

We believe that over time, the traditional customer segmentation approach will instead be replaced by personalized service spectrums and a "total service model". In the survey 92% supported the idea that *"Hotel guests will expect their stay to be personalized around a set of choices they make at the time of booking or prior to arrival"*. This will mean the guest of the future will be able to tailor every aspect of their experience including technology, hotel services, the bedroom, the journey, pricing and communications. Each guest will have their own preferences, demands and characteristics. The challenge for hotels is to understand and act upon these evolving requirements.

6   Technology Enabled Customization

Technology is helping to bring about this personalization of the travel experience. For example, via our smartphones, we can receive "augmented reality" digital overlays of information on real world objects to enhance our experience. Hence, I can now look at the Brandenburg Gate in Germany and with my phone scan around to see what the Berlin Wall would actually have looked like before it came down.

7   Personalizing the Hotel Experience

For hotel's personalization and effective use of technology are expected to be major brand differentiators – at least until the competition catches up or overtakes us! For example, social media is now a serious channel for hotels to build their brands and develop customer relationships. Hotels such as Joie de Vivre already use social media very effectively to build and reward a loyal fan base by making regular special offers to those registered on its Twitter and Facebook sites. In response to the rise of social media, 96% felt that *"Hotels will need to develop strong social media „listening skills" to understand how customer needs and perceptions of brands and service quality are truly evolving and to develop service propositions, marketing messages, and pricing solutions that reflect the needs of an increasingly diverse customer base"*.

The rise of social media also creates the opportunity for customers to define their requirements individually and collectively, and reverse auction them to the travel providers who can best respond. In the survey 90% of respondent agreed that *"Customers will increasingly use social media and collective intelligence travel services (like Dopplr) to define the desired „product" for a temporary self-forming group"*.

Technology will also enable us to capture and build up guest profiles that can be updated on each visit, based on what facilities and equipment the guest actually uses and does during their stay. Advances in areas such as artificial intelligence and modelling are also enabling us to become better at using customer data to anticipate future guest requirements. Perhaps unsurprisingly, 95% expect that *"Hotels will increasingly look to new technologies to drastically increase efficiency, reduce costs, personalize the customer experience and improve service"*. Despite the emphasis on technology, people will still be the ultimate differentiator – 93% agreed that *"Highly trained staff backed up by technology will be key to delivering personalized service and experiences"*.

8   Evolving the Intelligent Hotel

The potential to deepen customer insight with technology is almost limitless. For example, embedding cheap motion sensors in all the furniture and equipment in a room will enable hotels to monitor how often each item is used and for how long. Over time, as hotels are refurbished and new properties built, this data on usage and customer preferences would allow owners and operators to get a feel for what they actually need – rather than putting one of everything in every room. Accurate usage data and increasingly modular designs will thus enable hotels to reduce the total amount of furniture, fittings and guest equipment required for a hotel.

9   Delivering a Total Service Model

Over time there will be an entire spectrum of options over which the customer could be offered a choice. These range from selecting their own room, to deciding whether they want a TV or other audio-visual equipment, choosing the bed and linens, specifying the brand and range of amenities in the bathroom and picking the artwork displayed on the walls. For example, how many people would prefer to have their own photos or the work of their favourite artists displayed in a digital frame rather than somewhat forgettable artworks? In the case of TV's, while many people now bring their entertainment with them on their phone, tablet pc or laptop, there may be an interesting revenue opportunity in hiring out the latest technology such as 3D TV's or their successors to those guests that request them.

Hence a total service model means that you could have customers paying very different rates for adjacent rooms in the same hotel. One may simply want a good bed for six to eight hours sleep and a place to hang their clothes. Their neighbour may be looking to really embed themselves in a room for a few days and have a totally immersive environment. Of course there will always be customers who just want the standard offering, but there will also be this growing group who want this more personalized approach – selecting and paying for the elements they want. Expectations around personalization are high – 86% believe that "*By 2020, personalization will have been embraced wholeheartedly by the sector and that „customers will have the ability to choose the size of room, type of bed, amenities, audio-visual facilities, business equipment, etc. on booking and pay accordingly".*

Clearly there are concerns over the cost of moving to personalized service model and whether customers would bear the cost. The research suggests the demand will exist – with 92% agreeing that "*In a highly automated world, there will be a range of customers at every price point who are willing to pay for personal service".*

Technology will also allow us to transform the future guest experience. In the next few years we can expect to see augmented reality, room environment management systems and service robots become commonplace as guests demand even greater personalization, increased comfort and more innovative experiences. As we look to 2020, we will have to consider innovations such as intelligent furniture, personalized nutrition and responsive technologies that understand our cognitive functions.

The key to success in personalization lies in really engaging the customer through genuine dialogue about service improvement. We can have these dialogues during their stay, via open innovation processes, and by paying close attention to what's being said by guests on social media forums.

10 Beyond Hotels – The Rise of Virtual Experiences

The discussion so far has focused on how hotels can respond to evolving guest expectations and use technology to enhance, personalize and deepen the physical travel experience. However, increasingly, technology is also offering the potential to address the needs of those who want to experience a destination but without the time, cost or environmental impact of physical travel. This is the point where hoteliers need to close their eyes or cover their ears!

Already through webcams, virtual worlds, augmented reality, 3D virtual reality and other immersive technologies I can get a feel for a destination and travel experiences through the eyes of other visitors. For example, in soccer, the UK's Premier League is aiming to launch a 3D television service within the next five years that will give fans the experience of sitting in any part of the stadium they choose and watching the game as it would be experienced by those physically in attendance. Developments such as super wide photography and 3D graphical representations will be combined to recreate the live stadium experience.

11 From Virtual to Immersive

Advances in science and technology will continue to extend the potential of virtual experiences. For example, the cognitive sciences are constantly breaking new ground, teaching us more about the functioning of the brain and the electrical impulses that are triggered by each of our senses. Once these electrical patterns have been decoded, we will be in a position to go well beyond sharing just the audio-visual experience of a trip to the Galapagos Islands. The next stage in immersivity and augmented experiences will be to recreate electronically the smells, tactile experiences and taste sensations as if we were there for real. As we look at what is coming out of the labs, it is likely that such developments are only a matter of 5–10 years away.

The more immersive the technology becomes, the closer we will be able to mimic the live experience through virtual channels. For example, the Maldives could today recreate itself in a virtual word down to the last detail. In the near future, using multi-sensory virtual reality, it could provide virtual tourists with an experience something close to the physical one. Populating this virtual world with locals, hotels, restaurants, service staff and other visitors will help enhance the experience. New employment opportunities could be created for people to act as a hotel concierge or tour guide in the virtual world. The virtual holiday can also become a year round "any time any place" experience. Quiet Sunday afternoons, boring daily commutes on public transport and long train rides can all be transformed through immersive technology.

Virtual tours on demand will allow us to dip in and out of a travel experience at will. How different might you daily life become if you could tour the Taj Mahal on your way to work, stroll down the banks of the Ganges on your way home and join in the Mumbai Diwali celebrations after dinner?

12 Memory Transfer – The Ultimate in Immersive Reality?

With the developments described so far, the virtual traveller or "user" will still be able to distinguish between those vacations which they have experienced physically and those which they have only consumed electronically. However, there is the potential to extend the experience to "full memory transfer" – where it will be much harder to distinguish real from virtual. As science deepens our understanding of how we process information and encode our memories, so we are learning how to transfer electronic information directly to the human brain. Experiments have already been undertaken where individuals have transmitted numbers, colours and basic images to each other wirelessly.

The ultimate goal here is what inventor and futurist Ray Kurzweil terms "the singularity" – the point at which we can all connect to each other, share information and deepen our "collective intelligence" via the internet or its successor. Kurzweil argues that the exponential rate of development in fields such information technology, genetics and the cognitive sciences mean that the singularity could be with us as early as 2045.

13 Full Experience Transfer

Long before the singularity arrives, we should be in a position to capture every aspect of the multi-sensory experiences of a physical tourist, encode them and then enable others to download those experiences directly to their memories. At that point it really isn't clear whether we'll still be able to distinguish these new downloaded memories from those based on our own physical experiences. The potential then emerges to offer tiered pricing for different categories of experience. So would you prefer to see London through the eyes of the chauffer driven billionaire staying at the Savoy, or go for the experience of the backpacker staying in a youth hostel and duking it out with the rest of us on the subway?

The potential for disruption to the social order is immense. Imagine the scene, you walk into the office on Monday morning and begin to recount the amazing week you've spent in Venice with Megan Fox or Brad Pitt. To you horror and confusion, you discover that two other workmates are claiming to have had exactly the same experience down to the magical kiss in the gondola and fantastic spaghetti marinara in that quaint restaurant just off St Mark's Square!

14 Winning in Turbulent Times

So having explored the far horizon, let's bring the impact analysis closer to the present day. What seems clear is that we are entering an era where deep customer insight coupled with smart technologies should enable motivated, smart, well-trained employees to deliver a genuinely guest-centred service experience. While the words may all be familiar, the reality is that most hotels have some way to go to reach the kind of personalized service experience we've discussed here. So what will it take to make the transformational journey implied by the study's findings? In the report, we identify ten key characteristics of the successful global hotel brand of the future:

1. An organisation capable of surviving and thriving in turbulence and uncertainty

2. A portfolio of strategies for an evolving marketplace

3. Deep understanding of an increasingly geographically, financially, generationally and attitudinally diverse and rapidly evolving customer base

4. Delivering a personalised experience through a wide spectrum of service choice

5. Immersive, tactile and multi-dimensional technology interfaces

6. An open, listening, collaborative and experimental approach to innovation

7. Continuous search for ancillary revenues

8. Connected, adaptive and predictive

9. Asset light, insight rich

10. Continuous evolution – the hotel as a living laboratory

We explore four of these in more detail below.

14.1   A Portfolio of Strategies for an Evolving Marketplace

In the survey 81% felt that *"By 2020 hotels will increasingly experiment with a range of business models"*. The study highlights a range of alternative strategies and models that the branded hotel sector might consider in response to ever intensifying competition. These were tested in the survey, and the key findings were that by 2020:

- 78% agree that global hotel groups will increasingly seek to cover the full spectrum from budget through to luxury and heritage properties

- 79% expect that a new category of co-branded and co-designed "signature" properties will emerge within hotel chain portfolios, providing differentiation and opening up ancillary revenue stream options

- 77% think we will see the emergence of a new breed of unbranded hotel group, offering "white label solutions'- including sophisticated marketing, very high standards of service and advanced technology support while allowing owners to develop their own brands

- 46% anticipate the emergence of invitation-only hotels'

- 69% predict that hotel groups and owners may increasingly seek to co-locate different categories of hotel from budget to luxury in a common location with shared catering and leisure facilities for use by all guests.

14.2   Deep Understanding of a Rapidly Evolving Customer Base

Our customer base is becoming increasingly diverse, with three key drivers. Firstly, we all expect to see growth in visitors from developing markets as new middle classes emerge and start travelling for the first time, but it's by no means clear how profitable this business will be. Secondly, as the average age rises in developed economies, this is creating a growing group of older and often wealthier travellers who have differing needs to younger business or leisure guests.

Finally, the fallout from the global financial crisis means we could see an even broader spectrum of guest types and needs coming from more established markets – demanding shorter breaks and increasing value on the one hand and a more and more unique luxury experience on the other. Combined with what we've just discussed about personalization, these drivers suggest we will need to take our approach to customer insight to whole new level.

14.3   Immersive, Tactile and Multi-Dimensional Technology Interfaces

A key characteristic of tomorrow's successful hotels will be the ability to master and accommodate a rapidly evolving range of technology interfaces coming into our hotels in the hands of our guests and our staff. I have already talked about how augmented reality is blurring the boundaries between the physical and the digital world. At the same time, the hotel lounge of the future will be home to voice recognition, gesture interfaces, heads-up displays, projection screens, 3D displays, touchable holograms and an ever-widening array of interfaces through which we'll display and interact with our computers, phones and data.

14.4   Continuous Search for Ancillary Revenues

Confronted with economic uncertainty and intensifying competition, there is a clear need for hotels to invest time and energy in driving revenues. In the research 91% predicted that „*In the face of intense competition, hotels will increasingly turn their attention to generating ancillary revenues through activities such as increasing their share of the spend of each guest staying at their property*". Options might include:

- adopting alternative pricing mechanisms such as auctions of spare capacity, best price guarantees and "pay what you think it's worth';

- capturing a share of pre- and post-trip travel spend – 80% predict that hotels will use discount offers on the purchase of luggage, clothing, transportation, insurance and duty free;

- selling the products guests have experienced in their rooms – 67% expect that Hotels will create their own catalogues of branded amenities, clothing, furniture and decorations;

- business support services – 89% believe hotels will increasingly provide additional services e.g. translation, access to local legal and accounting advice, secretarial support, company formation and organization of small meetings.

15 Embracing New Management Thinking

The challenges and opportunities outlined in the report suggest that hotels will need to place an increasing emphasis on transformational strategies and on approaches to innovation which place the voice of the customer at the heart of the change agenda. Approaches such as open innovation and crowd sourcing are now in widespread use in other parts of the business world and the hotel sector will need to learn from best practice on how to truly value voices from outside the organisation and the industry. In the survey, 96% felt that „*In the face of intense global competition, the hotel industry will develop a strong focus on strategy and innovation – adopting approaches such as crowd sourcing and open innovation to generate new ideas*".

Hotels 2020 represents a wakeup call to the fact that the world has changed fundamentally. The path of the economy and hotel market over the next ten years is uncertain. So we have to prepare for a range of possible future scenarios. This implies development of leaders, managers and staff who are curious, tolerant of uncertainty, capable of scenario thinking and willing to make decisions with imperfect information.

16 Conclusion – Transforming Hotels from the Outside In

So, to conclude, hopefully this article and the report it's based on have highlighted that we are entering a world where the winning strategies of the past are no longer certain to succeed in a fast changing and turbulent world. Brands that don't recognise and respond to this run the risk of falling seriously behind the competition. While some of the developments discussed here are twenty years or more from affecting us, others are already having an impact and the pace of change is only likely to accelerate. Responding effectively means a fundamental reframing of how we view our hotels. We need to develop a mindset that enables us to rethink our strategies, revolutionize business models, rework service delivery and – as a result – reinvent the customer experience.

The transformational changes we are describing will require bold decisions, bold actions and bold new thinking. We can't rely on incremental improvements to marketing and service delivery to attract tomorrow's guest. We may have to redesign fundamentally the services we offer and re-imagine how we'll relate to an increasing diversity of future guests with constantly evolving needs. To succeed, hotel groups may increasingly view themselves as being in a constant state of experimentation – with the individual properties as living laboratories for the development and testing of new ideas. In "Hotel 202'0 every customer interaction could be viewed as a potential source of feedback, new ideas and competitor insight.

Hotels 2020 is not a destination but a transformational journey. This requires us to develop deep customer insight, embrace technologies that will help us enhance the visitor stay and develop leadership and a workforce than can go beyond segmentation to deliver a truly personalized guest experience.

Forecasting Global and European Tourism

Rolf Freitag and Dennis Pyka

1 Introduction

This Report is based primarily on the 2010 results of IPK International's World Travel Monitor – the continuous tourism monitoring system that was set up 23 years ago. IPK now conducts more than half a million representative interviews a year in 60 of the world's major outbound travel markets – 35 in Europe, 15 in Asia and 10 in the Americas – representing an estimated 90% of world outbound travel.

The interviews – more than 6 million of which have now been undertaken since 1988 – are designed to be comparable from one year and from one market to another, and to yield information on market volumes and sales turnover, destinations, travel behaviour, motivation and satisfaction, travellers and target groups, recent tourism trends, and short- to medium-term forecasts.

The report offers an overview of trends in world tourism, the European outbound market and finally, the report includes a summary of IPK International's view of the prospects for travel and tourism in 2011, based in large part on its Travel Intentions survey conducted at the start of the year.

2 Overview of World Tourism in 2010

2.1 Global Outbound Trends

In 2009 the industry came down to earth with a bump. The global economy is always subject to downturns, yet in recent years a downturn in the general economy meant, for the travel and tourism industry, merely a pause in growth. This time it was different. Not since the World Travel Monitor was launched have we seen a worldwide decline on this scale.

The world travel and tourism industry recovered in 2010 from the recession and is firmly back on the growth path as consumer demand for leisure travel improved.

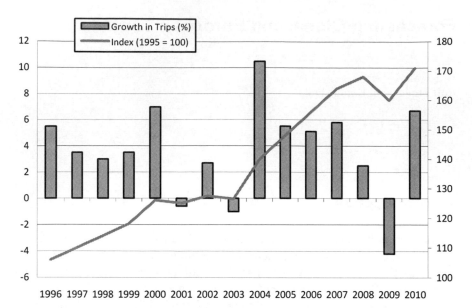

Fig. 1. World outbound tourism performance, 1995–2010

Source: World Travel Monitor, IPK International

The worldwide number of outbound trips recovered strongly in 2010 and rose 7% last year to 698 million after a 4% drop in 2009, according to results from IPK International's World Travel Monitor®. This is a new all-time high record in terms of trips.

The number of overnight stays went up to 5% to 5 billion. This was an improvement after a sharp drop in 2009 but the total remained below the 2008 level. Similarly, international travel spending increased 7% to €781 billion last year, which was a good recovery from 2009 but the spending figure remained below the all-time high record in 2008.

The total figures for arrivals cannot be strictly compared with the World Tourism Organization's (UNWTO's) estimated 935 million arrivals in 2010 – since this figure includes same-day trips and arrivals by children under the age of 15, as well as cumulative arrivals in several countries visited on one trip.

According to its own specific measure — which includes total spending related to a trip – the World Travel Monitor suggests that expenditure last year rose by 7% to €781 billion – €112 per night and €1,100 per outbound trip.

2.2   Inbound Overview

The international tourism recovered strongly in 2010 from the blow it suffered due to the global financial crisis and economic recession – faster than expected. That

% change in world arrivals by month in 2010

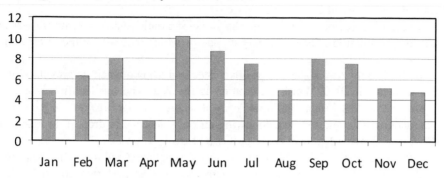

Fig. 2. World tourism performance by month, 2010

Sources: World Travel Monitor; World Tourism Organization (UNWTO)

means the vast majority of destinations worldwide documented positive growth in international tourist arrivals: all together almost 7% rise of the international tourist arrivals (935 million). In contrast, 2009 worldwide arrivals suffered an 8–10% decline (877 million).

However, the first part of 2010 dominated uncertainty based on the depressed growing rates of the previous year and the disappointing slowdown in April (only

% change in international arrivals

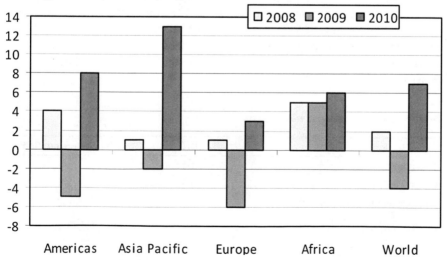

Fig. 3. Growth in international inbound tourism by region, 2008–2010

Note: Asia Pacific includes West Asia/Middle East and Central Asia

Sources: UNWTO; World Travel Monitor, IPK International

2% increase), when the ash clouds after the eruption of a volcano in Iceland caused the closure of European airspace.

The tourism sector now faces the challenge of a multi-speed growth – slower in advanced economies, faster in emerging ones – which will last for the foreseeable future.

In 2009, only Africa showed positive rates and it maintained its growth with a 6% increase in 2010. Worth mentioning is the Asia and the Pacific (including West Asia/Middle East and Central Asia): it recorded a significant 13% growth in international inbound tourism, following an only decent 2% drop. That makes it the fastest growing region over the past two years.

After a decline of 5% in 2009, the Americas rebounded and reached a new high in 2010 (8%). Only Europe showed again the weakest performance since 2007 with a 3% increase, following a 6% decline in 2009.

Few sub-regions departed far from the averages for their regions as a whole. Western Europe did relatively well, partly because of the strength of some economies in the region (notably Germany), and partly because some countries (especially Switzerland) were recovering from poor performances in earlier years.

3   European Tourism in 2010

3.1   Overall Travel Demand

Europeans went travelling again in greater numbers last year than in 2009. In 2010 European adults aged 15 years and over made 401 million trips abroad of a minimum one night's stay, according to IPK International's European Travel Monitor®. This represented an increase of 2% compared to 2009. The number of outbound room nights declined 1% to 3.4 billion, and the total spending travel increased by 1% to €337 billion.

Table 1. European outbound travel, 2010

	2010	% change on 2009
Trips (mn)	401	2
Overnights (mn)	3,395	−1
Average length of trip (nights)	9	−3
Spending (€ bn)	337	1
Spending per trip (€)	833	−1
Spending per night (€)	99	2

Source: European Travel Monitor, IPK International

3.2   Purpose and Length of Trip

Of the 401 million trips made in 2010, 287 million were for holidays (including short breaks), which generated 72% of all European outbound trip volume. This represents an increase of 3%. In spite of the much-publicised increase in social and employment mobility in the enlarged European Union, visits to friends and relations (VFR travel) have not been growing. Indeed, after declining by 10% in 2009, VFR trips were down a further 2% last year.

Table 2. Purpose of travel by Europeans, 2010

	Trips (mn)	% market share	% change 2010/09
Holiday	287	72	3
VFR & other leisure	56	14	–2
Business	58	14	–1
Total trips	**401**	**100**	**2**

Source: European Travel Monitor, IPK International

In terms of holiday times, Europeans went city-hopping last year, with the number of city breaks up strongly by 12%. The number of tours also grew well, by 7%, and the sun & beach travel, which dominates the outbound holiday market in Europe, also had to document a slight rise of 1%. Holidays in the countryside declined by 8%, however. Touring holidays, which had been doing exceptionally well in recent years, fell by 14% in 2009, but received a noticeable boost of 7% this year.

Table 3. European outbound holiday travel trends for selected segments, 2008–2010

Type of trip	% annual change		
	2008	2009	2010
Sun & beach	5	–3	1
City breaks & events	5	–3	12
Touring	14	–14	7
Countryside	2	–6	–8

Source: European Travel Monitor, IPK International

The number of trips of one to three nights – the most dynamic part of the European market in recent years – had risen by 7% in 2010, resulting in a slight increase in market share, to 29%. Meanwhile, the number of longer trips of four nights and more remained static by 285 million, or a share of 71%.

Table 4. Short breaks/trips versus long trips, 2010

	Total (mn)	% of trips	% change 2010/09
Short breaks (1–3 nights)	115	29	7
Long trips (4+ nights)	285	71	0
Total trips	**401**	**100**	2

Source: European Travel Monitor, IPK International

After a significant fall in nearly every mode of transportation for European outbound travel in 2009, it experienced an improvement in market shares. But there is still a 3% decline in low-cost airline travel although these low-cost airlines did manage to increase their share of the overall market. As well as travel by rail suffered a decline (5%).

The most surprising change is the travel by ship with a double-digit increase of 24%. Because prices for e.g. Cruises sink more and make it more possible for tourists to afford this.

Table 5. Mode of transport for European outbound travel, 2008–2010

	% annual growth		
	2008	2009	2010
Air	1	–8	2
– Low fare	4	–4	–3
– Other	–3	–9	0
Car	3	1	2
Coach	1	–19	3
Train	6	–6	–5
Ship	7	–15	24

Source: European Travel Monitor, IPK International

Table 6. Short- versus long-haul travel out of Europe, 2010

	Trips (mn)	% market share	% annual change		
			2008	2009	2010
Short-haul	352	88	2	–5	2
Long-haul	51	13	3	–10	4
Total trips	**401**	**100**	2	**–6**	2

Source: European Travel Monitor, IPK International

It is important to note that popular destinations in the southern Mediterranean – such as Egypt, Morocco and Tunisia – are counted as short-haul/intra-regional destinations out of Europe by World Travel Monitor, while in UNWTO's statistics they would count as long-haul destinations. By IPK's definition, short-haul trips account for as much as 88% of total outbound European trips.

3.3   Major Source Markets

There were ups and downs among the main source markets in Europe last year, the IPK figures showed. The European source market showing the biggest drop last year was Great Britain with a 4% fall while the German market declined by 1% in terms of the overall number of outbound trips. In contrast, Russia achieved a remarkable comeback, generating a 17% rise in outbound trips last year, after declining in 2009. Spain also grew strongly (4%) and there were slight increases of 1% for France, the Netherlands and Italy, as well.

(mn trips and annual percentage change)

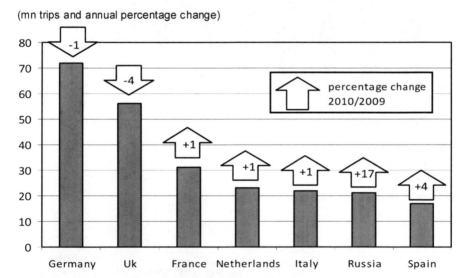

Fig. 4. Leading European outbound travel markets, 2010

Source: European Travel Monitor, IPK International

3.4   Major Destinations

The top ten destinations for European travellers are all within Europe, with the exception of the USA, in eighth place. This year, they fared nearly the same from one another in 2010: Germany, Italy, Austria, the UK, Turkey, the USA and Croatia recorded a significant increase in the ranking.

After a 1% decline of the leading destination of European outbound travellers Spain and a surprising 11% increase of the now second placed outbound market

Germany, the gap between those is smaller. The French market stagnated and hence, lost one place in the ranking from second to third place.

In contrast to the trend in the western Mediterranean – the tried and tested favourite European holiday destinations – countries on its eastern shores and on the Adriatic performed rather better. Croatia and Turkey actually showed a positive growth – Croatia reached a 5% rise and Turkey 3%. Only Greece recorded the highest drop of 3% in the leading destinations in 2010.

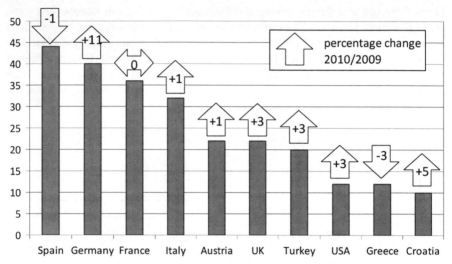

Fig. 5. Leading destinations of European outbound travellers, 2010

Source: European Travel Monitor, IPK International

Breakdown of trips and annual % growth in 2010

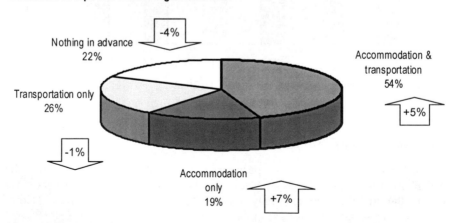

Fig. 6. Europeans' advance bookings for outbound trips, 2010

Source: European Travel Monitor, IPK International

3.5 Booking Patterns

In 2010, the "turn up and go" travel (with nothing booked in advance) recorded only a 4% decline after a significant drop of 20% the year before. There was also a slightly 1% decline for "transport only" pre-bookings (reflecting the search for bargain prices and an unwillingness to be surprised by prices at the airport). "Accommodation and transportation" pre-bookings rose by 5%, raising their market share from 41% in 2009 to 54% in 2010.

To bring up the rear, "accommodation only" pre-bookings with its significantly 7% increase have only 19% market share in 2010.

The proportion of holiday trips booked with the help of the internet exceeded those booked without for the first time in 2008, and the trend continued apace in 2009 and 2010. And the use of the internet for holiday trips continues to rise more: there is a 12% increase of bookers and 27% use the internet to search for information, but do actually make a booking. But the focus of growth is now firmly on holiday trips actually booked online.

Every year the number of holidaymakers not using the internet decline. In 2010 it is another 14% drop to a market share of only 37%.

Table 7. European online travel for holiday trips, 2008–2010

	% of holidaymakers			% increase 2010/09
	2008	**2009**	**2010**	
Use of the internet	56	60	63	15
Bookers	41	48	49	12
Lookers[a]	15	12	13	27
No use of internet	44	40	37	−14

[a] Use of the internet to research travel options, but not for booking

Source: European Travel Monitor, IPK International

4 Looking Forward

4.1 Economic Recovery

Although the financial crisis is not yet over, the world economy is recovering and the tourism industry will further benefit from this recovery. IPK expects global tourism growth of 3–4% in 2011, although the tourism world will remain divided. In the "new world" – the emerging markets – tourism demand will grow by 10% in Asia, 7% in Africa and 5% in Latin America. The "old world" – Europe and North America – is still in a phase of consolidation with modest growth rates of 2%. According to the latest surveys of IPK's World Travel Monitor® changes expected in the travel and booking behaviour. Thus, more off-season trips and further increase in internet bookings are expected. Social media and smart phone use as sources of information or help with actual travel planning will also increase.

Good prospects for the world economy in 2011 and beyond will be a major driver of growth for tourism. The world economy will grow 4.2% in 2011 and 4.7% in 2012, according to Germany's respected IFO Institute. "The world economy is on the brink of a consolidated upswing," Dr Gernot Nerb, IFO's director of industry research.

World economic growth is being driven by Asia (mostly China and India) and Latin America, the economics expert said. "Emerging countries will contribute about two thirds of world economic growth this year," Nerb said. In contrast, US GDP will grow 1.9% in 2011. Western Europe is improving but there are big differences from country to country. Germany is "a front-runner" with 2.2%.

4.2   Prospects for Travel and Tourism Demand

For the moment, European outbound travel is far short of its peak levels: trip volumes in 2010 were back to their level in 2007 and spending fell back to its 2004–2005 level. And, overall, there are no clear signs of a self-sustaining, long-term recovery in European outbound travel demand.

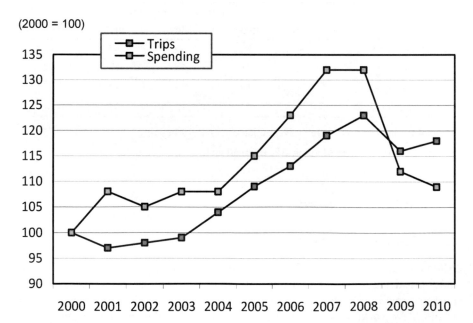

Fig. 7. European outbound travel and spending, 2000–2010

Source: European Travel Monitor, IPK International

As was the case through 2010, the World Travel Monitor's latest survey wave (carried out in January 2011) included a section on travel intentions over the coming 12 months. The sample base represented all those in 23 major countries of origin who had travelled abroad in 2010. The results suggest the situation is improving, but with widely differing prospects for the outbound markets of Europe, the Americas and Asia Pacific.

Asked whether the financial crisis is continuing to affect their travel behaviour, 46% of Europeans replied "No" and 25% said "Yes". By comparison, in September 2009, 52% replied "Yes". Of those who replied in the affirmative, 28% said they planned to take a less expensive trip; 30% said they would travel in off peak season, and only 16% admitted that they might not take a trip in 2011, planning instead to travel again in 2012.

Overall, IPK's European Travel Confidence Indicator for 2011 (where 100 would be neutral) stood at 102, although this average masks wide variations from one market to another.

(% of respondents)

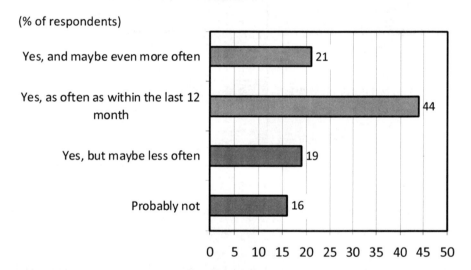

Fig. 8. European outbound travel intentions in 2011

Source: European Travel Monitor survey, January 2010

The Americas are also recovering well this year but with two very divergent trends. North America is only slowly picking up from the 2009 downturn but South America is growing fast, driven by a Brazilian boom.

Looking ahead to 2011, South Americans are more optimistic than North Americans in their travel planning, the Americas Travel Monitor shows.

The North Americas were asked the questions about the effect of the financial crisis on their travel behaviour, too. 58% answered "Yes" and 42% "No" – only slightly better than the 65% who replied "Yes" in September 2009. Of these, 38%

said they planned to take a less expensive trip and 24% that they would spend less money at the destination – much higher proportions than in Europe. Moreover, 23% said they would travel at home rather than abroad, and 26% that they would probably not travel in 2011.

Overall, IPK's Travel Confidence Indicator for North America is 98 – four points weaker than Europe's, but on a par with the European source country showing the lowest confidence level. IPK expects North American outbound travel to decline by 1% in 2011.

In contrast, IPK's Travel Confidence Indicator for South America is standing at a healthy 104. IPK expects South American outbound travel to increase by 5% in 2011.

(% of respondents)

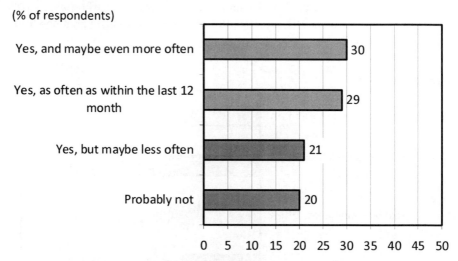

Fig. 9. South American outbound travel intentions in 2011

Source: World Travel Monitor survey, January 2010

Asia Pacific is demonstrating this year that it is one of the engines driving world tourism forwards. At present Asia only accounts for 18% of world outbound tourism, compared to 59% for Europe but is already ahead of the Americas (17%), according to the Asian Travel Monitor.

In January 81% of Asians, when asked whether the financial crisis would affect their behaviour, replied "No". The remaining 19% replied "Yes".

Some 19% of Asians said they planned to take a less expensive trip this year and 22% that they would spend less on food, shopping, etc. A further 15% said they would probably take a shorter trip, and 12% that they would not travel at all. IPK therefore puts its Travel Confidence Indicator for Asia at 108 – and somewhat higher in China – expecting Asian outbound travel to increase by 9% this year.

(% of respondents)

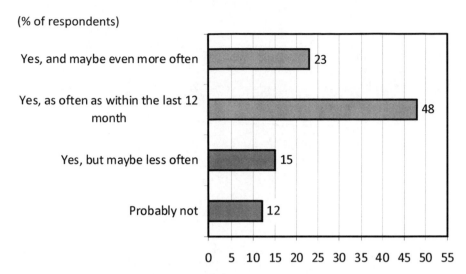

Fig. 10. Asian outbound travel intentions in 2011

Source: World Travel Monitor survey, January 2010

International tourism is now predicted to increase in the 3–4% range in 2011, according to IPK. IPK's Global Travel Confidence Indicator lies at 103 points for 2011, indicating 3% to 4% growth in world outbound travel. The main drivers will be the improving world economy and low fares from budget airlines while safety and environmental concerns are not expected to impact significantly.

Destination Management – Culture, Landscape, Cities

Mongol Passion: History and Challenges – Can Tourism Be a Tool to Empower It?

Damba Gantemur

When I studied in Europe, I had the chance to explain my belief regarding the meaning of Mongolia in relation to the meaning of humans and humanity. Once, my friends asked me "Gana, do most Mongolian people have similar almond eyes like you? Have you ever heard that Down syndrome was called Mongoloid disease, are there many people who have such syndrome in your country?" They wanted to know my reaction and were also interested what it means to be Mongol. We all knew that it was called such because the person who first characterized Downs as a distinct form of mental disability (John Langdon Down) believed that children with Downs resembled Asian people in facial characteristics, and so he called them "mongoloids". It is now considered a derogatory term, and rightly because there is not any medical or genital link between the Mongolian ethnic group/people and Down syndrome, it was renamed. So our discussion immediately went to the meaning of why or what does Mongolia mean. There are various hypothesizes about meaning of Mongol in Chinese, Japanese or ancient Mongolian sources, like it is the name of a barbaric tribe, a primitive belligerent or mon as name of river, mountain, tribe, togetherness etc. Also it means a silver-mountain (Mongon-Uul=Monggoul), or eternal mountain (Monkh-g-Uul=Monggoul). Another interesting hypothesis is Mongolia as it is derived from "Mun" which originates from "Hu-Mon", it means "mankind". In fact it is argued that Mon (Mongolia), Mun (Humun, Hunnu), Man (man /woman) all have the same origin, therefore Mongol (Mon+Uul as plural affix) means as "Men" or "Human" and "Humanity".

A Hunnu Empire which was the first powerful nomad state to appear in Central Asia in 209 BC also had a similar meaning of "Human" or "Man". Mongolia celebrates its 2220 year anniversary of Hunnu Empire in 2011 and many festivals will take place. The name of Atilla, the Hunnu (Northern Hun) king who led his men all the way to the walls of Rome, Hungary and the Hunnu Empire stretched from Baikal Lake in the north to Great Chinese Wall in the south, from the Yellow Sea to the oases of Central Asia. The state, ruled by a king or Shanyu elected by an assembly of all tribe chieftains- khurultai, was built on the principle of military democracy under which all the nomadic herders were warriors and subjects at the same time. Chinese historical records note that each autumn all men and cattle were counted to decide the amount of taxes and army subscripts. Recent research suggests that Hunnu did not differ much from modern Mongols in their appearance and may represent their ancestors. Anthropological studies show that the Mongoloid race or Central Asian type was already well shaped by the time of Hunnu. This a final conclusion made by Prof. G.Tumen, Chair of the Anthropology and Archeology of the Mongolian National University, after more than 30 years of comparative study of skulls from Stone Age to modern times. DNA analysis also proved the consistency of genetic lines between Hunnu and modern Mongols. This scientific conclusion implies that Atilla the Hun was indeed an ancestor of Chinggis Khaan.

For Mongols, who traditionally revered their ancestors, Chinggis Khaan is a God. His name and his doctrine has been perceived much deeper than just a national pride or hero, rather it means a perfect expression of solidarity, spiritual force, lodestar and the object of not only of national but of personal pride. But historiography Chinggis Khaan has been very biased and mainly written by the side that conquered him. The West did not have very high opinion about the Mongols, and viewed them as despotic, barbaric, and utterly cruel society as the West became wealthier and more sophisticated after 1240. Missionaries traveling to Mongolia generally did not report very favorably about them, and were unimpressed by the lifestyle and culture of the empire. In Central Europe, where Mongolian invasion in 1239–1242 caused some noteworthy destruction, to call someone a Mongol was a slur. In medieval historiography, Chinggis Khaan was viewed similarly to Attila as utterly barbaric, primitive, bloodthirsty, and evil. Several Moravian (Moravia was invaded in 1241) historians from 13th century onward describe Golden Hordes" khans as despotic, arrogant, vainglorious, and an alcoholic. This makes studying this time frame very difficult since best resources were collected during the medieval era, and present historians are left with modern interpretation. Russians were mainly considered as darkness of history under the name of Tatar/Mongols. In the Muslim world, Chinggis Khaan is depicted very negatively by historians with biased writings. Iranian writers are the most biased, thus almost all sources available in Afghanistan and that region are referred from those books. Therefore I suspect the general impression and historiography of Mongols was quite negative to the outside world.

Modern historians attempt to write more positive views based on our original historical sources and oral popular historical tradition regarding the Mongol empires, and nomadic Mongols to western literature. One of current best examples is, a book, "Genghis Khan and the Making of the Modern World" which was published in 2004 as a New York Times Best Seller by Jack Weatherford, Dewitt Wallace Professor of Anthropology at Macalester College. In 2006, he was awarded the Order of the Polar Star, Mongolia's highest national honor, and highly respected by our people indeed. It is not just he who has the ambition to present Chinggis Khaan in a far more positive light than traditional Western historiography; also he got in touch with the very internal spiritual world of Mongols regarding our history to the western people in the western logic. For example, he stated "Genghis Khan was an innovative leader, the first ruler in many conquered countries to put the power of law above his own power, encourages religious freedom, create public schools, grant diplomatic immunity, abolish torture, and institute free trade. The trade routes he created became lucrative pathways for commerce, but also for ideas, technologies, and expertise that transformed the way people lived". However, the most controversial statement in the book is the statement that the European Renaissance was a rebirth, not of Greece or Rome, but of ideas from the Mongol Empire as stated "Under the widespread influences from the paper and printing, gunpowder and firearms, and the spread of the navigational compass and other maritime equipment, Europeans experienced a Renaissance, literally a rebirth, but it was not the ancient world of Greece or Rome being reborn. It was the Mongol Empire, picked up, transferred, and adapted by the Europeans to their own needs and culture".

Mongolian scientists and historians avoided a true insight into the history about Chinggis Khaan, as it was too general or negative, and he/she had to arrest or victimize people should they have passed on positive written stories regarding him. My childhood and school age was during the time when Mongolia was in socialist doctrines, therefore almost all historical books, history even all arts, movies were written by Soviet ideology and socialist historiography. I can say that Mongolian people who are over 35 years old have been educated through socialist interpretation of knowledge especially on our history and religion. Everybody had studied books and courses which were created according to Marxist and Leninist philosophy, so Socialist Party controlled all sciences, thus, the whole society was propagandized as Mongolian monks, spiritual leaders were yellow feudal or black capitalist and Chinggis khaan was conqueror and killer, which put him to shame.

Now an interesting question mark has arisen to me. Is our current respect and love for Chinggis Khaan just a commodity or show that in the case of most of our adult people in Mongolia have been educated to be shameful of him? Or has our internal soul been revived again since we chose to be a democratic country in 1990? I strongly believe that a "Mongol passion" has been existent in the pastoral nomad life which contains the conventional wisdom about the world, their deep historical pride, spiritual power, and an ecological adaptation that makes it possible to support more people in the Mongolian environment than would be true under any other mode of subsistence. Mongolia is a birthplace of ancient nomads who played a prominent role in their time not only in the political history of the whole of Central Asia but in its spiritual life as well. The Mongol Passion has been flourishing time to time, was rooted from Hunnu Empire and shaped an actual wisdom during the Great Mongol Empire. Unfortunately the historiography was really very biased, so there has always been a misunderstanding gap between Mongols and outside world, thus, our passion has been interpreted by others. It is definitely not an easy question to discuss in one small article. But I would like to share my opinion of it in briefly. After Ikh (Great) Mongol Empire, which was from Chinggis Khaan's declaration in 1206 till to the death in 1370 of the last emperor of China's Yuan dynasty, Mongols often faced prejudices from the outside world and faced several political and socio-economical challenges which came directly from the outside.

The first one was probably during the XYII-XX centuries, when Mongolia was under the Manchu domination. In the early XYII century Central Asia, as the Mongol empire disintegrated, the Manchu emerged. After the hundred-and thirty year long endeavor, Manchu conquered Mongolia, so darkest period in the history of Mongolia was the 275 years of Manchu domination. By that time Mongolia had been divided into three parts: Inner Mongolia, Khalkh and Oirad. Manchu began carrying out administrative, military and economic reforms. Dividing Mongolia into numerous very small units was sure to aggravate Mongols" tendency toward individualism and independency. Especially Manchu/Beijing had a very careful preventive policy themselves from revival power of Mongols. Bawden noted that "the Manchus were quite uninterested in the economic or social development of Mongolia," using it instead mainly as a buffer with the Russian Empire. But Manchu was not able to break the internal soul and will of Mongols which was fostered by plenty of stories, myths and narratives of Chinggis Khaan. Therefore his name was not just disseminated as a name of person; rather it was perceived motto and passion of freedom & independency. The Mongols has been inspired by their passion more for than two hundred years, and on the December 29, 1911 Bogd Khaan of Mongol state was founded. 2011 is the 100 year anniversary of the Mongolian independency day.

Manchu very much supported to develop Tibetan Buddhism in order to lessen the power and courage of Mongols, but it has had quite an opposite effect which the Manchus were unable to foresee. As matter of fact the more Mongols became Buddhist the more it spiritually estranged itself from the alien conquerors. Most people thought Mongolian Buddhism is just a copy of Tibetan Buddhism and Lamaism. It is predominantly the Yellow Hat sect of Buddhism practiced in Tibet and China. But it has evolved into its own version, having incorporated the pre-Buddhist religion of tangarism as well as shamanistic influences. Most importantly, Buddhism has harmonized with nomadic way of circumstances which is people on horseback freely galloping in open steppe, living in the Ger in very harsh climate between -40 +40 degrees. During the Manchu time, Mongolian identity was reinforced by spiritually and also intellectually under the Buddhist superficiality. Mongolia became one of the three highest peaks of Buddhist Civilization during the Manchu time. Mongol nobles and monks created millions of works on the Buddhist "five great" and "five small" sciences, all of them were grounded in the essential part of Mongolian culture which are expressed from our daily habits custom, events till philosophical view even nowadays. So Mongolian culture and spiritual soul was not only fostered during the Manchu time, but it also enriched the Buddhist culture and practiced in every step of nomad life. That is why Mongols still kept their own uniqueness during the next challenged time.

The Socialism was next period to be challenged, which tried to assimilate Mongols" will. The socialism continued for almost 70 years, from Mongolian spiritual leaders, monks and high educated people who had shown their struggle and tried to keep their national pride and Mongol identity against Soviet controlled socio-economic development. Socialism had provided a particular regime of truth. Therefore political repression is one of the sensitive topics of Mongolian modern history and it occurred throughout socialist period. Although political motivated killings began in the early 1920s and reportedly continued until 1985, the period of greatest repression was the late 1930s. Victims (around 100,000) came from all levels of society, many of them (around 20,000) were Buddhist monks, lamas who were enlightened monks, spiritual leaders and others were political and academic figures, nobility and ordinary workers, herders were also included. It was called the Great Purge, which was launched in 1937, and Moscow played core role in it. At the end of 1939, a whole population of Mongolia was only 700,000 and about 20 percent of adult people were killed and more than 60,000 people arrested during that time. After the Second World War, there was strong pressure to follow the Soviet model in socializing the economy and society instruction. The status quo of Mongolia was accepted at the Yalta Treaty which was a result of unremitted revolution since 1921. On October, 1945, based on the returns of referendum, China and Soviet Union officially recognized the Mongolia and we became officially independent, Nation-State after more than forty years of fight for independence. That time, the propaganda was that we were at end of the anti-imperialist and anti-feudal stage of the Mongolian revolution, so now we had to begin building the foundation of socialism. The new socialist way of life, collectivism and agriculture was started up. Following the soviets example a campaign of collectivization (Negdel, Kolhoz) was organized once again in 1960 after first attempt failed in the 1930s. In fact animals were taken away from the herdsmen by force. A negdel was organized into several brigades that were mostly nomadic. The members of a negdel received wages and were entitled to holidays and pensions. Dependent on the geographical location, herders were allowed to keep 10–15 private animals per family member, but no more than 50–75 per family. This system tried to alter a Mongolian mentality which a herdsmen can be a proprietor and decider of his own life, live in surrounded Ger that always encourage a concession instead of tension, into a "Homo Sovieticus" thinking and a characteristics which is always subsidized by others, depended on others. However, during the socialist time, there were three well-organized cultural attacks on health and education. Especially skin and venereal diseases were eliminated, illiteracy was abolished and an urban way of life was introduced.

I want to argue that nowadays Mongolia faces another challenging period. Mongolia, unlike other Asian Transitional economies, has since 1990 pursued a "shock-therapy" or "big bang" transition to a market economy and democratic policy. The country moved into deep recession from 1992–2000 with the collapse of the command economy, as the country eased price controls, liberalized domestic and international trade, and started the restructuring of the banking, energy and tourism sectors besides livestock and mining. Because Mongolia had just opened to the world, the inbound tourism has rapidly developed since 1992, attracting travelers through its unspoiled nature, nomad life, proud history and unique wildlife in general. The general mainstream tourists quite like to discover Mongolia; additionally niche-market special interests such as horse riding, fly-fishing, cultural expeditions, and nomad based community tourists have increased. Related with Mongolia's location, neighborhood countries, South Korea, some of Western European countries, Japan are the main markets of inbound tourism. The attitude of Mongols regarding the way of tourism development has been showed in three stages since 1990. In 1990–1998, the international tourism was developed by itself in a way of tourist facilities and services were established in the areas which are popular places for Mongols to visit. People very much keen to do tourism business even if it has a strong dependency on seasonality is why tourist camps and other facilities had mushroomed. Tourism has seen as a friendship form with the international world, a bridge between various cultures, an expression of foreign commodity and goods, an important source to foreign exchange currency and a mode of internationalization. The main perception of tourism was that a tourist is interested in everything that Mongolia has to offer, therefore, we just send them to the countryside in summer time and were able to use all the income in winter.

Since 1998, facing an enormous amount of economic and social problems to the transition period, inflation has already increased, poverty is growing in the city with small-middle scale businessmen keen to have a real estate as a guarantee in case of future risks. So tourism as stated is one of the priority sectors of socio-economic development, thus several policy papers have completed. The main goal is set, we have to build many constructions and complexes in the unspoiled places in order to attract as many tourists as possible. So in the most known places fences, cement camps, complexes are now being built. I believe it is a consequence of the socialist way of thinking, because it seems very familiar to me as socialism propagandized a development which means to build a material and technical base of socialism in the virgin lands. In the last decades, Mongols also travelled a lot to the foreign countries and tried to create similar constructions from their experiences of what they have seen from settled touristic areas based on the tourism service complexes in mass tourism attractions. Therefore the value chain process needs to be improved and the role of tourism has to be defined as a socio-economic environment which is developed by the country. In particular we should look at what the core experience of Mongolia is for the western market. How to sustain the core value of Mongolia which is not only beauty landscape and wildlife but also a living cultural heritage related with today's nomads" tradition in particular not only what you are able to see but what you can feel from this country.

At the moment the attitude towards tourism is quite interesting during this process. Most of us could see, the economic value of the mining sector so the role of tourism like other sectors significantly decreasing, especially tourism due to the development of infrastructure and buildings instead of socio-cultural and ecological values.

The remarkable thing is that Mongolian economy is fully dependent on natural resource sectors, thus sustainability is a core issue of tourism sector in Mongolia.

Economic activity in Mongolia has traditionally been based on herding and agri-culture-Mongolia's extensive mineral deposits, however, have attracted foreign investors. The country holds copper, gold, coal, oil, molybdenum, and fluorspar, uranium, tin, REM and tungsten deposits; thus, mining sector accounts a large part of foreign direct investment and government revenue. Our economy continues to be heavily influenced by our neighbors, we purchases 95% of our petroleum prod-ucts, a substantial amount of electric power from Russia, leaving it vulnerable to price increases. Trade with China represents more than half of Mongolia's total external trade- so China receives about two thirds of Mongolia's exports. We joined the World Trade Organization in 1997 and seek to expand its participation in regional economic regime. Recent years, Mongolian economy has hugely de-pended on few main products such as copper, coal and gold. Around 70% of the Mongolian export goes to China and our main product prices hugely depend on the Chinese economy. Raw materials of animals, such as wool, leather and cash-mere almost all go to China and sheep and cow meat tried to be exported into China and Russia, probably to Gulf countries. In October, 2009 Mongolian Par-liament signed the "Oyu Tolgoi" Investment agreement, it has amazingly rich with copper resources, which are estimated at 32 million tons of copper and 1000 tons of gold. Experts say this number will increase. This one of the largest untapped copper deposit, scheduled to begin in production in 2013. The mine will account more than 30% of the Mongolian GDP. A "Tavan Tolgoi", which is another strategic important mining deposit, holds about 6 billion metric tons of coal (1.8 billion metric tons of high quality coking coal) in the southern Mongo-lian desert, making it one of largest unexploited fuel reserves. It is expected to begin production in 2012.

Soviet assistance, which managed to amount to one third of the GDP, disap-peared in 1990, since then more than 100 recognized international development organizations have implemented various projects in different fields, especially from Japan, EU, World Bank, ADB, USAID, GIZ, IMF therefore the foreign donor organizations have been very important for the Mongolian government due to their loan power and aid to this country. Since Mongolia stated that mining would be the main economic sector of Mongolian development, most international organizations were eager to leave Mongolia and even the World Bank expressed that Mongolia is as a "middle income country". In fact the gap between the very rich and poor people is quite critical especially in the city; Mongols have to think about our future in terms of green development and green economy. Because we are on the verge of development, I would like to remind about our previous ex-periences of Manchu and socialism also how western world misunderstood us in terms of historiography. It is to be noted that many years of foreign domination did not turn the Mongolian passion into an appendage of their culture into which Manchu and Soviet Bloc themselves were in the end assimilated.

I very much believed that tourism especially the western market is highly im-portant to the future development of Mongolia. But it is not just foreigners that teach what sustainability of tourism to nomadic civilization is, particularly the fundamental principles of sustainable development indeed fits well into the nomadic lifestyle. Probably most of us see tourism development as something with which to achieve a similar level of city tourism, therefore McDonalds, Burger King, or Starbucks should be established here. Unlike other countries, core experiences of Mongolian tourism are not directly connected to monuments or certain sites but to the rural nomadic population. They are not site-specific, which is why competitiveness between destinations inside Mongolia to a large extent depends on the ability of local people to develop skills that meet tourist needs. Mongolia's tourism resources directly consist of and depend on local people, local knowledge and their assistance, which provides what tour operators have promised to their clients. Foreign tourists, especially Europeans seek to experience the authenticity of Mongols and the meaning beyond nomadic people and unspoiled nature. A nomad friendly tourism strategy fosters sustainable development by inspiring nomads" traditional connection with their herds and pastures. Particularly, tourism economically empowers small middle scale entrepreneurs in rural places and herder based livelihood. Not only avoiding Dutch disease, but also keeping our core values and uniqueness, the sustainable tourism development should be strategic role in case of mining and cashmere (goat) sectors predominated in natural resources throughout Mongolia. Mongols are very much respectful people especially to their guests, you can see our hospitality if you go to the countryside. It enhances the core mobile identity of Mongols which I already called the "Mongol passion". Because I believe sustainability of tourism is not only achieved from supply side, it is also very much depended on the demand side and mutual cooperation.

The 10 Brand:Trust Theses on the Future of Alpine Destination Management/Branding

Klaus-Dieter Koch

Even though the principles and methods transforming touristic destinations into highly attractive brand systems (destination branding) are still relatively young, the changes in value such as fragmentation of the target groups, orientation towards sustainability, radical new communication and media models and the drastic economy changes, have forced us to rethink present convictions and to explore new marketing paths.

The following theses were developed on the occasion of the Destination Days at the ITB Congress 2011 in Berlin. They are based on the comprehensive project experience in creating effective destination brands, the intensive exchange with CEOs, brand managers and academics and the analysis of future challenges. They are to serve as inspiration and orientation to create a proactive process of change.

* * *

The hour of marketing strikes whenever an oversaturation caused by too many and too similar offerings threatens markets by causing a fall in value and prices. Since the beginning of modern mass tourism, market-oriented supply, price, distribution and communication politics have created a differentiated, inexpensive and attractive offer and have thereby elevated worldwide tourism to an unprecedented level of professionalism.

However, it is exactly this professionalization which has caused the applied instruments to be overused. Over the years the customer has got used to a lasting "thrill" und therefore constantly demands the new and unknown. What was once the journey over the mountain pass Brenner to "Bella Italia" in the 50s, the holiday package to Majorca in the 70s and the low-cost flight to the Dominican Republic in the 90s is now a self-awareness stay in a Burmese cloister and tomorrow perhaps a flight into space with Sir Richards" space shuttle. The approach to find true needs as is taught by marketing is almost exhausted and its instruments are becoming less effective. More of the same will lead to a higher interchangeability and thereby to dramatic fall in prices.

The brand system offers development, a superordinate performance and significance which orients itself less on short-term and easily reproduced needs and instead more on "life shortages" of people. The shortages in life are wishes, dreams, desires and hopes and accompany people all their lives. To combine and express these ideally suited response fields with own performance superiority is the aim of brand management.

This cause-and- effect relationship is confronted by special demands destinations have. The main element of building a brand is superior performance. To secure and express this is the main task of a destination brand manager. Compared with companies, independent service providers face an especially difficult challenge due to their multitude and variety. To motivate, coordinate and monitor these, a homogeneous, consistent experience must be provided which places high demands on organization and on those responsible at all points where the customer comes in contact with the brand. Historically, the Alpine region has one of the longest experiences in tourism worldwide. Therefore, intensive use is made of an extremely sensitive ecological and social system. The strong growth the region saw in the past is hardly possible there in the future. Still many tourism managers" performance is measured by the number of overnight stays and bed occupancy rate. For economic reasons this is surely understandable but with limited space resources and an already strained infrastructure this is no model for touristic future. Should a fall in prices occur due to competition of other destinations then new investments will no longer be possible and the appeal of a destination will sooner or later decrease, which would result in a further price drop. The existence of mainly medium-sized companies will be threatened due to stricter lending criteria (Basel IV).

What are future potentials of Alpine tourism? Future potentials which harmonize the complicated demands of an experienced customer with the economic necessities of service providers and the transparent, yet still different global competition. We also don't know whether the following assumptions will occur as predicted and especially don't know when. However we and our business partners know one thing for sure – change will become necessary and has to be actively shaped by those responsible.

1. The yardstick for future success is no longer the number of overnight stays but added value per guest.

Pure volume growth can no longer be the sole yardstick for success with the Alpine region having limited resources. The Alpine experience needs space, scenery, intact nature, culture and intact social structures on site.

For the Alpine tourism the only way is to move up to the premium segment in each guest category (ranging from school and youth trips, singles and families to the "silver generation"). The aim has to be to offer first-class service which has a high reference and thereby gain appropriate premiums.

2. A widely-known destination is no longer important, but its attractiveness is.

Being widely known in a chaotic market with overlapping poorly defined structures is no longer of value and is only a requirement to gain access to a competition of attractiveness. Purchase decisions are constantly being made on a short-term basis, more spontaneously and with less commitment, which is why the destination brand will be favoured, because it manages to develop the highest attractiveness in the long-run in its respective competitive environment.

3. To suit everyone's taste leads to nothing, that's why less is more – intense and extraordinary.

When comparing tourist services in Alpine destinations it becomes apparent that all have something to offer everyone. Whether in summer or winter, the basic offers are generally the same, exchangeable and by now trends don't differ due to the high level of networking. However, to stand out from the majority of providers, specific services are required which ideally are linked to the tradition, history and culture of the location and can only be found there.

4. "Old school" tourism marketing wanting to create illusions doesn't have a future role to play, instead clearly defined and believable high-end services at all points of contact have to form a well-balanced experience package for the customer.

The future guest will have more experience, knowledge and access to information than ever before. Classic advertising and marketing concepts are no longer convincing. Direct information from a provider is mistrusted. Even Tripadvisor is being superseded by Facebook because the recommendation comes from a personally known source (=friend) and enjoys a far higher reputation than the unknown trip advisor (which could also be manipulated).

Focus on needs ("what does the guest want?") is no longer the center point of branding, instead credibility is ("does he want it from me?"), based on demonstrable high-end services.

5. In the future, the guest is searching for more sense and is no longer satisfied with superficial offers obviously created for him. Background information of a destination and not the touristic offer will become the decisive factor.

The guest is looking more and more for background information about a destination. Information on history, culture, agronomy, nature, tradition, social life etc. suggest a very specific experience to the guest who is searching for meaning and makes the destination that manages to do this unique and uncopyable.

6. Away from territory, on to themes.

Destinations and their management are territory-related. The region or the location is the focal point. However, the guest's purchasing decisions are less dependent on the region but more theme-oriented. He doesn't travel to Basel but to the world's largest art exhibition. He doesn't drive to Tirol but to the world-famous Hahnenkamm Race. It's no longer about marketing a certain region, valley or location but about creating themes and connecting these with the destination in a believable and lasting way.

7. Ability to cooperate and network within a destination as the key to future success.

The guest expects an authentic product which should offer him a well-balanced experience package. To achieve this all service providers in a destination have to pledge to keep mandatory minimum standards within an agreed framework. This is where cooperation and good networking need to be developed far beyond the known concepts of items like the ski pass or mobility card.

8. Sustainability becomes taken for granted.

Everybody is talking about sustainability, but it is already no longer suitable for an effective image of a destination due to long previous investment periods. The guest in the Alpine region will simply expect this. In Tirol breakfast milk coming from Hungary is no longer tolerated and attractiveness is lowered by the exploitation of natural resources or landscapes over-filled with technological hardware. Therefore, sustainability is not to be seen as illusion marketing but as a future standard to be achieved. In future things won't work without this.

9. Digitalization wherever possible.

Digitalization makes everything faster and more direct, but also more non-binding. Classic processes in marketing, sales and performance are being turned upside down. Tourism manager tasks will completely change (who still needs tourist information if an app is available?). People's work in production will alter and at the same time become more significant for overall success. Standard performances and processes won't generate money in the future. Thereby entire components of the touristic value-adding chain will become redundant (e.g. all non-specialized agents).

10. Dynamic pricing and dynamic packaging turn conventional price and segment structures on their heads.

Pricewise everything will change. Now and even more so in the future. Therefore, prices will no longer function as an indicator (cheap=good) but become the biggest amplifier of mistrust (has my neighbor paid less than me?). Special emphasis will be placed on the question "what is it really worth?" and influence making decisions. Consequently, consistent value creation management of a destination will become decisive for the success of brand management.

City Tourism – The New Magic of the Place

Burkhard Kieker

"We must endeavour to remain as surprising as we are now."

Berlin's popularity as a tourist destination is increasing day by day. 30 million overnight stays in 2020 – this is an entirely achievable goal. The tourism boom of the past few months suggests that Berlin exerts an unusual power of attraction, an enticing magnetism.

I should, however, play down in advance any possible expectations that I, as the Head of the Destination Berlin marketing machine, am able to explain the reason for this success. I cannot explain it. On the contrary: even we are surprised sometimes. This would perhaps suggest that there is definitely an element of magic at play when it comes to what sums up cities and their power of attraction.

There are few statistics available which hint at the source of Berlin's magical power. We have just received one such statistic and it was a complete surprise, again. The January figures for Berlin came from the Berlin-Brandenburg Office for Statistics. And behold, in January, normally the nightmare month for every tourist destination, Berlin achieved a plus again: 6.9% more visitors, 4.6% more overnight stays – and these figures are for a city in the far north of the country, during a freezing, wintry month. We are, of course, delighted about this and hope that Berlin's growth trajectory will continue in future and that Berlin's magical power will long remain in force.

Last year saw a double digit growth. Prior to that, it was the year of the financial crisis, when things went steeply downhill for many of our friends and colleagues working in city tourism around the world. Berlin remained completely unaffected by this drop, experiencing an increase of 6.2% in 2009 (see fig.1).

Time after time, we ourselves are amazed by these statistics and have to ask ourselves where this magic could be coming from. I would first like to answer a different question: where do we actually come from?

Berlin is a city that, if we're honest, actually disappeared from the map. There was always much talk of Berlin in politics; the powers of the Cold War wrangled over this pawn, according to which historical perspective one took. As a tourist city, however, Berlin had disappeared from view. The city did, in fact, appear in a few spy movies, or people drove to the Berlin Wall and climbed on top, or gazed,

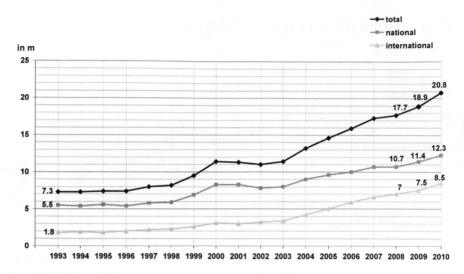

Fig. 1. Overnight stays 1993–2010

Source: Statistical office Berlin-Brandenburg (Berlin, 2011)

shocked, from the other side and were not allowed out – nevertheless, Berlin had disappeared!

A city that was the epitome of the cosmopolitan metropolis in the 1920s but which had since then simply bid farewell to world history. Bomb-ravaged, fragmented, oppressed, even the bourgeois elite who sustain a city were got rid of. The elite could not remain here; there was nothing for them to do. West Berlin was mostly full of young people: those who had either run away from army conscription, or who wanted to experience the Bohemian atmosphere in Kreuzberg – that summed it up.

When the Berlin Wall came down in 1990 everyone was euphoric – myself too, incidentally – and we thought: OK, let's get started! Nothing started.

It seemed only natural that, in 1990 for example, Japan Airlines established a non-stop flight route from Berlin to Tokyo, Malaysia Airlines thought it necessary to connect Berlin to Kuala Lumpur, and even Air Canada flew from Berlin to Toronto. All three did this for six months and then stopped. Why was this? Because Berlin was a grievously wounded city and in my view and in the view of many of Berlin's inhabitants, had taken a long time to recover from these wounds from sixty years previous. That period lasted for 10, 15 years and could not be transcended in just a few months. So much for the sad part of the history. Now for the good part.

What happened next is the history that you will read everywhere nowadays: Berlin as a city that everyone wants to visit, a creative city – a concept that has been used to such an extent that I now only approach it very carefully and with forceps.

Berlin – a place where plenty of exceptional people with good ideas and interesting life stories would like to live.

For me Berlin has in some ways emerged from the depths like the lost city of Atlantis. Many spoke of it. The city must have existed at one time and now it – we – exist again and many people come and visit us and want to see what it is like here. At the same time, we are still unripe, still unfinished. Staying with the Atlantis metaphor: there are still algae everywhere, we are still dripping, but we are here and we exist again.

This leads us to one of the "magic" factors that I would like to reveal: the city's unfinished nature. Naturally we, together with our colleagues at Berlin Partner, speculate on what Berlin's core brand really is. Following extensive discussions with numerous experts, we came to the following conclusion: **Berlin is a city of change.** Nothing that is here today will look the same tomorrow, everything is in a constant state of flux and this fascinates the majority of people.

Berlin has, however, yet another important charismatic feature. I am thereby trying to pinpoint the exact thing that we know virtually nothing about. We do however know what others say about us: journalists, foreign correspondents, visitors – they all claim to find Berlin authentic. As an inhabitant of Berlin, I cannot and should not assert such a thing myself. But others say this about us. The city is authentic. I believe that in this idea we are getting close to one of the factors of success for the future.

Berlin is authentic and authentic means that we are wild, we are a little disorganised, a little disordered, but life is never boring here. There are lots of people who are looking for exactly that. I do not believe that people look for something finished, set in stone, unchanged for 200 years, with no more surprises in store. At the same time, this is the challenge for our city, too. We must endeavour to remain as surprising as we are now.

What happened in the past? What were or are the requirements for Berlin's success as a tourist destination?

One of the significant requirements was, in my mind, that there was no marketing mastermind. It would be great if there had been one person, or maybe even an organisation, which would have sat down in 1990 and said this is what we must become, for everyone to like it here. There was no individual and no organisation which did this, though. Berlin even actively avoided this; it was a city that would not put up with it.

What originated here, originated via a chaotic process of ongoing change and evolution. On the basis of our figures, the only thing that we can say is that the result of this process goes down well with people. This is proven by almost 21 million overnight stays over the past year (see fig.2).

People are coming to us in droves. In some instances, they are even coming from markets where we, visitBerlin, have not previously invested millions in any large-scale, sophisticated advertising campaigns.

I would like to cite Italy as an example here: in Italy we have focused almost exclusively on PR. We invited a large number of journalists and relied on word of

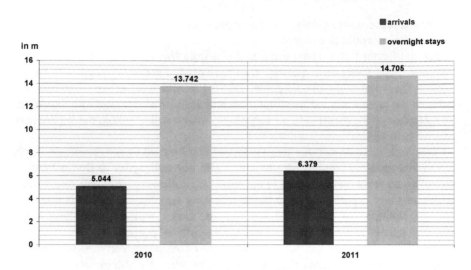

Fig. 2. Growth over previous year, January to July 2011

Source: Statistical office Berlin-Brandenburg (Berlin, 2011)

mouth to spread the word. This strategy was very successful: our visitor figures last year showed the Italians in top place (see fig.4).

Berlin can also boast another significant advantage: **Berlin is good value for money.** This does not mean that the city is poor. On the contrary, just take a drive through Grunewald – this will bear clear witness to the fact that the capital city is anything but poor. Berlin is a city that is good value for money for visitors. We hope that this too, will remain the case for a few more years to come. In Berlin, you can spend a weekend enjoying everything city life has to offer. This means everything that falls under the heading of culture – from the highbrow to the low-brow, from subculture to night life. You can, moreover, get the full experience here for half the price of New York or London (see fig.4). This too, has got about and is also a powerful attraction for tourists.

Prior to this, something different has to happen though: the city and its inhabi-tants must themselves be interesting. This is where the idea of the unfinished comes into play again. The fashion critic Suzy Menkes called it the "raw energy of Berlin". This is what attracts lots of interesting people.

Everyone who is anyone in the cultural scene wants to rent an apartment here. Artists, designers, musicians – they all come here, live here and form a critical mass of people who create an atmosphere in the city that is so inspiring that, again, other people come to see this atmosphere for themselves.

On top of that, these people are in a position to live and rent a studio here. Just try to rent out a studio in Paris with actual daylight when you are a young artist who has never sold a picture – this is possible in Berlin.

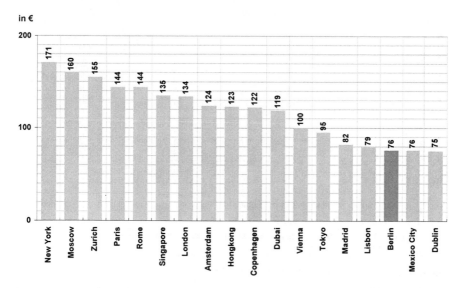

Fig. 3. Average room rates, first half-year of 2011
Source: Hotel Price Index, www.hotels.com (2011)

Berlin is, of course, also a city full of things to see and do. That's not the point here however. Most visitors know exactly what Berlin has to offer: "more museums than rainy days" – which is true, by the way.

Our visitors do not come here primarily for this reason. They come to Berlin to see lifestyles. Here again we find the connection with the theme of the "magic of the place" and also with the people who live in this city. Visitors come from all over the world to recharge.

Personally, I come from the province. I come from Gummersbach, Bergisches Land, so I know what I'm talking about. I know from experience that people drive to the capital – it will be exactly the same in other countries too – simply to find out how their country works. How do things work today? Above all: how will things work tomorrow?

This is a very important task for a capital city. It is also a very important point which promotes Berlin, since Germany had no capital city for 50 years.

60% of our visitors are native, i.e. they come from Germany, and 40% are our neighbours, or come from further afield (see fig.5) – this figure is still too low in our view. We would gladly switch these percentages around and, in future, receive more international visitors.

German residents come primarily to see: how does the Republic work, what can I learn about the true nature of Germany? We also have more and more international visitors, who come to share in the Berlin lifestyle.

All we can do is collect all these points, pieces of information, and try to fit them into the overall puzzle, this little work of art that is Berlin.

in m

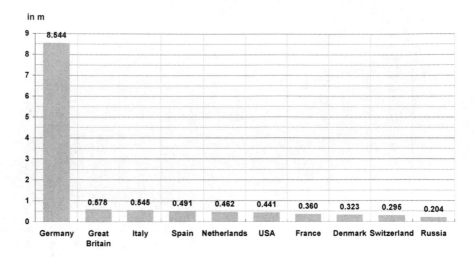

Fig. 4. Top 10 markets/overnights Jan–Aug 2011

Source: Statistical office Berlin-Brandenburg (Berlin, 2011)

Collaboration with the Foreign Office provides us with another important piece of the puzzle: at presentations – in Australia, New York or other places, too – colleagues always commented on Berlin being the most effective destroyer of stereotypes about Germany. Why is this? Because Berlin is a city where people work on their laptops in cafés. Because in summer people relax in beach bars. Lastly, because Berlin is an incredibly relaxed city in any case. Where it used to be said that the Germans would get up at six, start building Volkswagens and had absolutely no sense of humour, Berlin manages to refute all of these prejudices. I think that the people of Berlin like playing this role, too.

I have already referred to it: a magnetism that works by people influencing people. People come here, they live here, and they shape the city. In turn this leads to lifestyles being put on show.

The magic of the place also accounts for another critical factor in Berlin: Berlin is a historical focal point. Our national and international visitors know that a great deal has happened here – for better or worse. They therefore want to see exactly where things took place. When we say: "You've seen it in Hollywood, but it happened here!", then that is part of the magic of this place.

We work very consciously with this in close collaboration with the many museums, memorials and commemorative sites. However we also work, for example, with an organisation like the video bus tour, which tracks down films. Whether it is the famous DDR film "Traces of Stones" or the recently released "Unknown Identity". What a brilliant film. Liam Neeson takes the starring role and Berlin stars likewise alongside Liam Neeson. All filmed here. There is that "raw energy" again. It is only a car chase film, but has a clever plot. This has been the no. 1 box office hit in the USA for the last ten days – invaluable advertising for us. This is because it addresses the "melting pot" theme – being the easternmost city of the West, the westernmost city of the East. Berlin also appears as a place of science, or as a yawning chasm, it all plays a part. House fronts show off shiny V2A metal; in backyards you can still see bullet holes from the Second World War – this is authenticity. This interests people.

In conclusion, I would like to look at one final "magical characteristic" of Berlin. I was first prompted to consider it when friends from Paris recently observed that: Berlin is so beautifully non-capitalist. Non-capitalist? What does that mean? This means that the people in Berlin – and many find this a very good thing – do not talk much about money. The people of Berlin do not define themselves in terms of money. Here, it is more important that you can tell a good joke, or somehow present yourself as a sort of minor work of art, instead of driving up in a Porsche, for instance. In actual fact, you will find that there are not very many Porsches here.

The magic of the place – what does that mean in a time when, thanks to the internet, actual locations are becoming less and less important? If we sit in front of the box, we are of course living in a completely "dis-locationed" world. It is possible to visit any place in the world at any time in the day of course, take web cams or documentaries. You can go back in time, open up Wikipedia and get everything virtually, it is all secondhand information. We are convinced that the pendulum is swinging back and that this development per se triggers a deep desire in people for authenticity, for place, for self-localisation. Where am I? Who am I? Here I sit now on the bank of the Spree and look at the shore, at the National Gallery, which I visit later and say: I am here.

I believe that this is the kind of magic we in Berlin should play with more; then the visitors will keep on coming.

Liveable City – Sustainable Quality of Life as Success Driver for Urban Branding

Andreas Reiter

1 City Competition

International competition of cities is becoming tougher; city destination networking is growing denser (especially due to the growing point-to-point connections of low-cost carriers). The above-average growth of city tourism has led and leads quite often to a rapid increase in bed capacity in the big cities and thereby to high utilisation pressure.

Tourism experts – not least because of this – are facing the challenge to create tight packages with finely tuned attention strategies to attract the guest interest. Very intense competition demands a continuous development of the brand and – its underlying process – the product "city".

Cities are highly complex brand "personalities" and diverse social areas. In their long European history they have always embodied more variety than simplicity; cities are colourful sociotopes with a differentiated environment and functions mix. Diversity (variety) is *the* urban quality per se, this is also reflected in the basic components of (touristic) urban adventure economy, the four A's:

- Attractions
- Amenities
- Access
- Authenticity

It is only through a special composition that these components provide a lasting success of a city with its visitors and encourage them to return.

In the highly competitive global environment the necessity of differentiation of cities is growing. Profiling of cities such as:

- creative cities (Berlin, Barcelona, Liverpool, Copenhagen ...),
- cities of culture (Vienna, Salzburg, Aix-en-Provence, Bilbao ...),
- cities of science (Cambridge, Münster ...),

- cities of sports (Valencia, Vancouver, Innsbruck …),

- history teller (Rome, Prague, Venice …) and so on

has been combined with an ever remarkable brand oriented product development in the past few years (for example in the case of Valencia and its stringent sports events).

The specific image of a city establishes itself in the minds of consumers through brand management and branding – the brand as is well known is the reputation which precedes the city. The hereby created images are strong resonance fields which trigger emotional expectations in the human-being. On the part of the city this constantly requires a content and product oriented development and differentiation of their strategic occupied profiles. A clear topic leadership has to be the strategic goal.

City consumption is adventure consumption, and this always wants to be adapted to shifting lifestyle preferences and a changing set of values of the urbanites. Especially announced cities, the "Cool Cities", have constantly gone through their attractions management – they continuously include narrative adventure areas as well as hip exclusive locations.

As exciting as this may sound, the "conventional" urban attractions arsenal rarely develops a sustainable differentiation. The stylish landmarks of the global "star architects" (museums, opera houses, brand lands etc.) are too often interchangeable, the urban lounges, waterscapes, city-beaches etc. are too similar.

Repetitive product semantics (Cool Spaces) have led to many cities being homogenized as far as possible and to a "haven't we already seen this in …?" dismissal by professional city-nomads. In the meantime the "glory" effect (the elation of the "majestic" in urban design as per dramatic adviser Christian Mikunda) has become the basic factor, yet not the differentiation factor of a city destination. Visitors from far away won't be lured anymore by the twenty-seventh city-beach, especially not those who would secretly be welcomed (the multiplier and lead user). Also as sights and remaining city icons are pull-factors and only act as a magnet for first time visitors (who would visit Versailles a second time apart from hardcore historians?); new attractions have constantly got to be developed beyond the mainstream.

Usually city trips underlie diversified visitor motives which form the core, where inspiration and sensual immersion into a different, vibrant city life lie. Maximum vibes in minimum time. Vibes cannot be controlled; however they can be bought sometimes, for example with a ticket for the M10, Berlin's party tram. This tram offers the probably highest dosage of club varieties for party lovers – whoever takes the tram from Friedrichshain via the Prenzlauer Berg to the Nordbahnhof, can experience a cosmos of the best clubs in the city all around.

City consumers want to be surprised; they would like to experience meetings and adventures in a new unusual context beyond their ordinary everyday lives.

"People nowadays aren't visiting a city for its urban history or cultural specialties, but because of the sensory enhancement, which it promises."[1]

2 Liveable Cities

The question is which senses should be enhanced and in which way? The spiral of
superlatives –higher, faster, and shriller – eventually twists in vain. The operating
system of the experience economy – that is unshakeable – will be newly drafted.
When positioning cities the "basic melody" is currently shifting: where as in the
last few years up until now you would come across the *Creative City* as the strategic focus of *Urban Branding*, the future will be especially about the leading theme
that is "sustainable life quality", the *Liveable City*. The city as a lifestyle melting
pot will be utilized in a new way – with new actors and a new story.

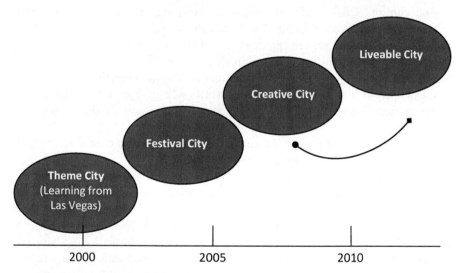

Fig. 1. Urban branding: from "festivalisation" to quality of life

Source: ZTB Zukunftsbüro.

2.1 The Location Factor "Quality of Life"

In the meanwhile the "quality of life" factor has become somewhat of a hidden
reserve currency in tomorrow's locational competition. Competition amongst
cities for talents, investors and tourists is becoming fiercer. As well as traditional

---

[1] Source: Design2context, meine, deine schönste Stadt der Welt. Merkmale urbaner Lebensqualität (2010).

hard locational factors such as economic performance, spending power, tax burden, infrastructure etc. the integrated city branding battle for attention is becoming more about the soft competition factor "life quality".

A city, worth living in, demands an ambitious strategy which targets optimisation of urban infrastructures and development of future-oriented subjects. The accentuation meets the inhabitants and tourists needs: visitors will only feel happy where locals also do.

City life quality is now being rated internationally, by numerous rankings (such as Mercer, Economist, Cushman & Wakefield amongst others as well as lifestyle magazines like Monocle). These rankings are used by cities as an important marketing tool and also for benchmarking despite of their debatable logic of evaluation.

As yet there is no binding scientific definition for "quality of life", although a methodical gathering from a mix of quantitative and qualitative factors has crystallized in the major international surveys on this topic.

The current "Quality of Life" research recognises life quality as a multi-dimensional cross-section factor. The term can indeed be defined with concrete variables and operationalized although life quality is rated subjectively (every human-being defines quality of life individually from his/her own living environment). The subjective well-being is determined by many objective values in situ (for example on-site living costs, social surroundings, educational and leisure offerings etc.). Thus a highly perceived quality of life is an indicator for high life satisfaction.

Life quality of a location "is the sum of external, objective conditions, subjectively perceived life satisfaction and own well-being."[2]

The results of a study carried out by BAK Basel Economics show that the strategic significance of addressing and acquiring highly-qualified people increases when occupants generally find life quality an important factor for a destination. In times when cities can only be successful when employing a concerted market appearance (this means don't utilise tourism marketing here and location or city marketing over there), stringent communication – and consequently also coordinated communication of the central profile topics – are the only effective way.

Quality of life may not seem like a sexy enough issue for a skilled marketing manager – it is however appropriate for an operating system of a city and brilliant for brand communication. After all each city has its own spirit, magic, its very own form of life quality and this can, should, must accompany the core brand. Thereby quality of life will become a central pillar of "competitive identity".

---

[2]   Source: Harald Pechlaner/Elisa Innerhofer/Monika Bachinger, Standortmanagement und Lebensqualität in Harald Pechlaner/Monika Bachinger (Hrsg.), Lebensqualität und Standortattraktivität (2010).

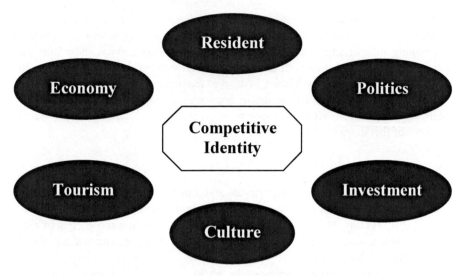

Fig. 2. Competitive identity

Source: Following Simon Anholt in WTO/ETC, Handbook on Destination Branding (2009)

Zurich – Downtown Switzerland: this claim compresses the brand essence (and thereby quality of life) of the Swiss metropolis: the international city in the Swiss mountains. The antipode city nature and metropolis countryside gently come together, contrasts seemingly unite in a lively and unusual way. Precisely this demonstrates the fine art of marketing such a complex entity which a city is.

The increasing importance of the factor "life quality" for place making and place branding doesn't come out of nowhere but from the slipstream of a shift in values in our western society. New wealth criteria, which aren't just geared to the GDP, are being discussed for some time now in all of Europe by politicians and economists. Eurostat would like to introduce important feel good parameters (e.g. sense of security) still in 2011 and also social variables (number of friends, club activities etc.) as additional values to the GDP in the future. The Better Life Index was implemented in spring 2011 by the OECD as a complementary measured value to the GDP. Wealth becomes well-being.

2.2   Strengths of a Liveable City

Sustainable quality of life – in its economic, ecological und social entanglement – is an interdisciplinary matter which considers all three target groups of a city (inhabitants, tourists, businesses). Visitors only feel well where locals also do. The strategic direction into the future is clear: the "feel good" factor of a city will become a central location factor with an ever growing accentuation of attributes such as "sustainable" and "eco-friendly". The city worth living in demands an integrated urban setting und thereby also a joint product development by the city protagonists.

An appealing city does not only offer a differentiated experience portfolio to its visitors and inhabitants, but also an excellent place to stay – which includes a functioning infrastructure, a supple systematic organised service chain and with increasing importance a refreshing interaction between residents and visitors. The city worth living in is all embracing instead of exclusive.

Desires and wishes of consumers change and demand an altered experience dramatic composition. Authentic communities interaction, green lifestyle are the platform of this dramaturgy.

Live like a Local

Modern performers, willing to experiment and urbanites with travel experience are searching for the true spirit of a city, the authentic encounters with locals, the "live like a local" feeling, they see themselves as locals for a limited period of time. This means that tourist experts have to form affinity groups and attract appropriate visitors over the lived values.

The future belongs to the communities where locals and visitors come together for a limited period of time. A good example is the Hotel Cosmo in Berlin. They have understood: guest and hosts are affinity groups, which have the same spirit and thus chose each other. "COSMO Hotel Berlin knows the people and places which keep Berlin moving. The concierge is their connection to the local scene and recommends announced restaurants, bars and clubs. He knows where Berlin's art and culture scene dwell. He sends them to aspiring city designers to go shopping …"[3] The "cosmos" of the hotel community appears on the hotels website: a variety of galleries, clubs, boutiques, bars etc. in the hotel's periphery. Same spirit, same station.

Interaction with Public Space

Exploring new urban milieus – away from the touristic ant trail and preferred in back yards – the vibrating microcosm city keeping the motto: detours increase local knowledge. Modern urbanites interact with public space; movements such as "Urban Knitting" (for example telephone boxes, lanterns etc. are covered in knitted wool overnight) testify this. What began in US cities and from there on (discreetly) spread in the Anglo-Saxon world doesn't have a lot to do with making places prettier rather than show the needs of people to shape their city (if only on a very small scale). Be it street art (by now probably described as an established form of art) or artistic performances like "skip conversion" (here skips are turned into swimming pools, table tennis tables etc. by young people) – the public space is being converted.

This temporary utilisation – it began a few years ago with the pop-up stores from the fashion industry – picks up an important character from lead users; they

---

[3]   www.cosmo-berlin.de.

make use of the city, they (re-)produce public leeway. This of course works primarily for one's own self but attracts like-minded people over the medium term. After all, these unusual city areas are detected by a new generation of city consumers where they can also unfold themselves experimentally.

The most legendary example has to be the Eisbach in Munich. The torrent amidst the Englischer Garten (English Garden) is still good for unique adrenalin kick, even though by now it is already an icon: young surfers made use of a raging point of the Eisbach to surf waves (illegally of course) many years ago, the reputation of this hot spot then spread rapidly worldwide among the surfer community, the scene tucked the surfboard under its arm and headed for Munich. Especially those non-constructed but grown experience spaces are also a part of the urban quality of life.

Trendy sports such as free running, street boulder, buildering amongst others are manifestations of an experimental acquisition by young people. Some climb bridges others house walls others jump across garage roofs. These performances are an expression of specific lifestyles (and by the way a great viral marketing tool for the city when communicating over Web 2.0 and clips) and belong to the lifestyle puzzle. Somebody who only sees behavioural problems in it doesn't understand city life.

The wind has changed. The cat walk was still predominant in the 2000s, the orchestration of the public area as a stage (Street Parade, CSD and much more) – nowadays the city area mutates more and more into a three dimensional space. For such freedoms cities have to hold out enough leeway (mental and juridical) and especially show good will politics to urban subcultures. Creating a young image and attracting young talents, be it as tourists or residents, can only be achieved with openness towards new trends. The unheard-of, the inappropriate will sooner or later rise to the surface.

Green City

Competition pressure is also increasing on cities and metropolis to introduce sustainable development strategies and to position themselves as cities with a small ecological footprint due to climate protection (cities are the main source of greenhouse gases) and consumption of resources.

The positioning as a liveable city demands making city living space more attractive under sustainable aspects; it does however go far beyond the green "marketing arsenal" (such as green spaces, green buildings, e-mobility, green meetings etc.). The mix of acceleration and deceleration zones, of urbanity and nature has to be communicated and developed strategically. Cities most worth living in (from Munich over Vancouver and Sydney to Copenhagen, from Vienna to Berlin) have these assets.

Lots of cities are already in the middle of an ecological restructuring, where growth is decoupled from resource consumption. Sustainability will be examined more often in the future by urban topics, from energy supply over mobility to the neighbourhood garden. Guerrilla Gardening is nowadays announced especially in the German metropolis (from Berlin to Munich), Vertical Gardens (by famous garden-artist Patrick Blanc) are flourishing in cities from Madrid to Vienna.

Fact is that the city worth living is the green lifestyles playground. New customer values (consumption with a good conscience) aren't just found in trendsetters. The green lifestyle also globally manifests itself increasingly in new unusual attractions – after all sustainability does have an aesthetic footprint.

The green High Line in New York as a new, green inner-city catwalk is such a distinctive symbol, or also the upcoming incinerator in Copenhagen which is combined with a ski slope and glowing CO^2 pillars of smoke designed by the BIG architects" office. These are the (green) hot-spots of tomorrow in city tourism – they are symbols of sustainable quality of life and are suitable for storytelling. Marketing means telling stories following the rules of the attention economy. Whoever produces the right symbols at the right time controls the market.

A new generation of urbanites is searching for the specific feel good factor of a city, which mixes creativity with green lifestyle, local authenticity with sustainable innovation.

Marketing and Distribution

The Needs of Package Tourists and Travel Agents – Neuromarketing in the Tourism Sector

Ingo Markgraf, David Scheffer, and Johanna Pulkenat

This article will examine, what makes package tourists spend a lot of money, compared to their day-to-day expenditures, on a service package of which true value appears weeks if not months after being purchased. Furthermore it will take a closer look at the motivational world of the travel agency staff and how both groups interact. These questions will be analysed with regard to its significance and applicability in brand management. The results of a neuropsychological study, which measured the implicit personality systems of package tourists and travel agents with a psychological test, are the foundations of this article.

1 Introduction

Only in recent decades, that holiday trips have been a possible leisure activity which is regularly affordable while still covering essential expenses (Mund, 2006, p. 115). Important drivers are more leisure time and a national income that has grown over the years.

For years sun, beach and relaxation have been the most important reasons for going on holiday according to travellers. Yet also other forms of holiday trips such as city, activity/sport, cruise or wellness holidays are enjoying an increasing popularity.

When investigating the travel market it must be taken into consideration that it is subject to considerable changes due to, for example, new dynamic production processes, price comparison systems, the growth of online providers etc.

2 Starting Position, Problems and Objectives

A brand core analysis in 2008 showed that brand packages of REWE Touristik GmbH, ITS, Jahn Reisen and Tjaereborg where being perceived as barely differentiated and hardly distinguishable. This applied to both customers and travel agents. The decision to continue the three brands led to the task of a new positioning, i.e. to make each brand unique and distinguishable in its perception.

The objective of the study which was carried out was, to better understand the package tourist market and come up with instructions for communication with package tourists through REWE Touristik brands. The key issues discussed where:

- Why do package tourists buy?
- Which scopes and potentials are there when positioning single brands?
- How can potential customers be better addressed and won as a customer?

Other issues where to better understand the travel agent group und to find out about their implicit effective sales motivation.

Central question concerning travel agents where:

- How do travel agents "tick"?
- What is their main motivation (commissions, incentives …) for selling a holiday?
- How can travel agents be addressed more effectively?
- How can travel agents help to increase the sale of REWE Touristik products?

3   Solutions and Method

Focus groups, interviews or questionnaires are conventional ways of finding out which needs are behind the behaviour of individual people and groups. Methods which record implicit needs and motives are used in rare cases if at all. It has been proven, that emotional and cognitive processes which lead to purchase decisions are at least 95% unconscious (Zaltman, 2003). This is why people can only give inadequate information about the real reasons for their decisions. This fundamental knowledge has been scientifically beyond dispute for more than 30 years (Nisbett & Wilson, 1977). However this knowledge is still being taken less into account in market research, marketing and brand management.

The implicit test "Visual Questionnaire" (ViQ) has been used to come up with the results used in this article. The test measures implicit personality systems objectively, reliably and validly by using an easy, purely visual method (Scheffer & Loerwald, 2008). The implicit personality systems as scales of the ViQ mainly describe how people perceive things and how they make (purchase) decisions (Figure 1).

The scale order in Figure 1 follows the functions of the personality system. Extraversion versus introversion are complementary ways of energy input and thereby an orientation on perception. Sensing versus intuition concerns perception itself, thinking versus feeling are decision strategies based on perception and judging versus perceiving relate to the handling of these decisions. The scales E/I as well as J/P are measured bipolar, S, N, T and F are all independent scales, which form the personality core area for perception and decision making. The abbreviation N for Intuition comes from an international convention, as the I already stands for introversion.

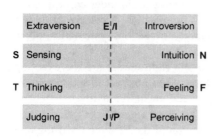

1. **Where do energy and motivation originate?** – from the outside (E) or the inside (I)?

2. **How is information processed?** - attention to detail (S) or with view to the wider context (N)?

3. **According to which criteria are decisions made?** – thinking (T) or feeling (F)?

4. **What is the lifestyle?** – straight (J) or flexible and spontaneous (P)?

Fig. 1. Explanation of the implicit ViQ personality system

For simplicity's sake the ViQ scales are shown in the two-dimensional "NeurolPS®-Map" system and referred to elsewhere as dimensions. The dimensions of sensing (S), intuition (N), thinking (T) and feeling (F) are combined to provide a four-element breakdown and together with the Extraversion/Introversion (E/I) and Judging/Perceiving (J/P) dimensions form 16 groups (Figure 2).

Figure 2 shows four key neuropsychological types, which are driven by different motives and needs. This is why different tonalities, lines of argument and triggers of emotions lead to the types feeling called upon.

Such customer types can be found in the top left corner of the NeurolPS®-Map, for which the classic economy image of homo oeconomicus generally applies, even though facts show that decisions are in the end based on an emotion – the

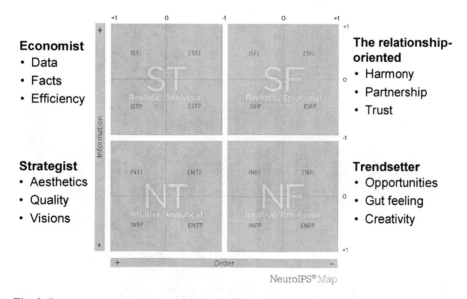

Fig. 2. Four core personalities and 16 types of NeurolPS®-Map

good feeling to be able to do everything right and have it all under control. ST or analyst types need a rational, clear cost-benefit ratio which is clearly evident to them to be satisfied. They have a need for reliability, precision, systematics, objectivity and efficiency.

Customer groups which are found in the bottom right corner are the exact opposite. NF – or Trendsetter types reach top form when abstract ideas, arts or the unconventional are involved. They have a high need for individuality independent of the price. Idealism for them is regarded higher than a good cost-benefit ratio, which is important to analysts. The basic motivation of the SF-type or of the relationship person is, just as the name states, relationships and harmony. The highest priority for them is a harmonic, empathic relationship to people and animals. Yet additionally SF-Types think practically and realistically. Diametrically opposite of the SF-Type stands the NT- or strategist type. Thinking about major visions and strategies are important to him. Logic and self-determination are valued.

Finally there are naturally also lots of hybrids, which are taken into account by using a graphical presentation of the Neurol®PS-dimensions in the analysis of package tourists and travel agents.

4   Study Design

A market research panel was commissioned to get package-holiday tourists to take part. 1082 package-holiday tourists of different brands took part (Figure 3), who clicked their way through a ViQ online and answered a few explicit questions at the end about brand and travel preferences.

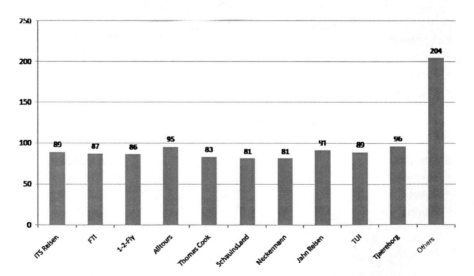

Fig. 3. Random number (German: "Teilnehmerzahl", English: number of participants) travel brands for package tourists, panel, N = 1082

An animated banner was designed with an editorial tag "What kind of travel-sales type are you?" to get travel agents to participate. The banner was put up on different travel agent portals. Ultimately 130 travel agents took part in the study (Figure 4).

Fig. 4. Animated banner to get travel agents to participate, what kind of travel-sales type are you?

5   Results

5.1   Needs of Package Tourists

Age and sex distribution of random samples was approximately the same as the nation's average.

The ViQ mean profile of all package-traveler samples allows a conclusion of the implicit personality systems and the needs and motives which are involved (Figure 5). Mean profiles for sub-groups can be created in detail (for example users of a certain brand) by analysing sub-group samples separately.

The ViQ mean profile is always worked out when the group of package tourists or one of their sub-groups is comparable to a representative norm of the total population.

The results are presented in z-standardised scales. The x-axis with the value 0 in the following graph shows a representative norm based on the mean of the population. The bars show the deviation of package-tourists or brand users respectively from this norm in z-standardised units. A value of 0.50 would therefore mean that brand users show a higher manifestation of one half of the standard deviation than the comparative norm. This would already equate to a very strong deviation or effect size.

Figure 5 illustrates that package tourists as an entire group only distinguish themselves in some personality systems by approx. one quarter of the standard deviation from the norm.

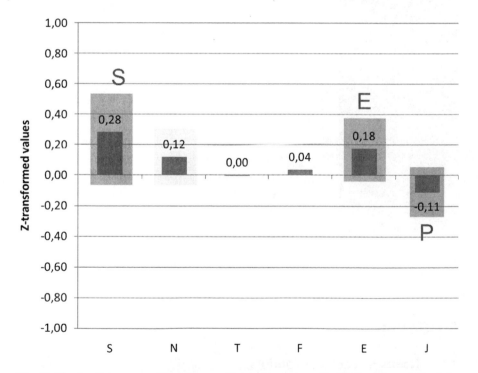

Fig. 5. The implicit personality systems of package tourists (N=1082) as the standard deviation from the German norm (N=100,000)

1. Higher values for the attribute "Extraversion": Package tourists pull more energy by interacting with the environment than the average of the population. They have a higher need for exterior stimuli and react quickly. They openly share thoughts and feelings.

2. Higher values for the attribute "Sensing": Package tourists have a higher need for realism and details. They rely more on their five senses for perception and less on their intuition. Believing means seeing for sensing types.

3. Higher values for the attribute "Perceiving": Package tourists are more flexible than the average of the population. They have the need to keep alternatives open for themselves and to make spontaneous decisions, preferably just before a deadline, based on perception of different perspectives. The here and now is more important to them than rigidly pursuing goals. This for example becomes clear when seeing the results of an additional question which was explicitly posed after the ViQ, whether a particular travel agent would be preferred. 50.7 percent stated that no particular travel agent is preferred. The brand TUI finished with max. 8.5%, followed by subsiding percentages of the brands Alltours (7.6 percent) and Neckermann (5.1 percent).

5.2 Description and Motivation of Package Tourists in Accordance with Their Types

Basically package tourists are an ESFP-type. EP-types are enterprising, always prepared to explore the new, lively, often loaded with energy, improvising and testing. They would prefer to savour the moment as if it didn't have any prehistory and implications. ESFPs as a core target group are open-minded, friendly and enthusiastic. They love to live (in a regulated framework) a colourful, rich and intensive life. Their behaviour is often shaped by their need for variety and interpersonal harmony. Impressions exceeding the norm appeal to them. Holidays in foreign countries offer a good opportunity to escape from everyday life, this being one of the most fundamental motives to travel.

In terms of the T and F decision-making dimensions package tourists are shown to be average in their nature. Package tourists find it relatively hard to make a decision. This is why package tourists happily accept the offer where the decision making and organization of a holiday is done for them. This convenience is far more important to them than individuality. In the extreme, carefree packages are the most significant part of services. When on holiday, package tourists want to have fun, "hang out", go on safe adventure trips and provide their senses with information – see new things, taste, touch, smell and hear. If a trip doesn't fulfill expectations then package tourists are able to improvise through the developed dimension P. This is why, compared to their everyday expenses, they are prepared to pay a large sum of money for a service package, the true value of which only becomes apparent on arrival at the holiday destination.

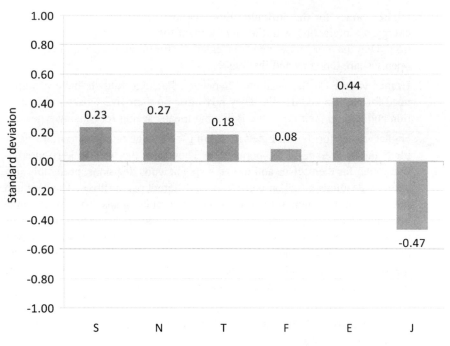

Fig. 6. The implicit customer personality systems of the brand Jahn Reisen (N=90) as a standard deviation from the German norm (N=100,000)

Of course the measured profile of package tourists is only valid in general for the overall sample. These profiles do partly deviate for customer groups of different brands. To this end Figure 6 shows the customer profiles of Jahn Reisen.

Jahn Reisen customers have stronger extraversion and perceiving attributes than any other package tourists. Additionally Jahn Reisen customers are clearly more intuitive than the average of the population and the other package tourists, i.e. they perceive the world abstractly and symbolically instead of realistically. Compared with other package tourists a higher degree of rationality is also significant. (Package tourists in total T=0.00, Jahn Reisen T=0.18). Basically Jahn Reisen customers, as a sub-group of the random sample of package tourists, generally belong to the ENTP-type.

This shows that Jahn Reisen customers are demanding, quick, innovative, imaginative, have wide interests, hands-on and love to keep in touch with the times.

Life seems to them like a dazzling spectrum of possibilities. Individual details are not so important. They prefer to look at the bigger picture with all possibilities. They mostly look towards the future. If given the opportunity Jahn Reisen customers prefer to consume exclusiveness – even if it costs significantly more. Logic is more important to Jahn Reisen customers than feeling. They feel especially well if they are consulted by independent, competent people who solve problems for them.

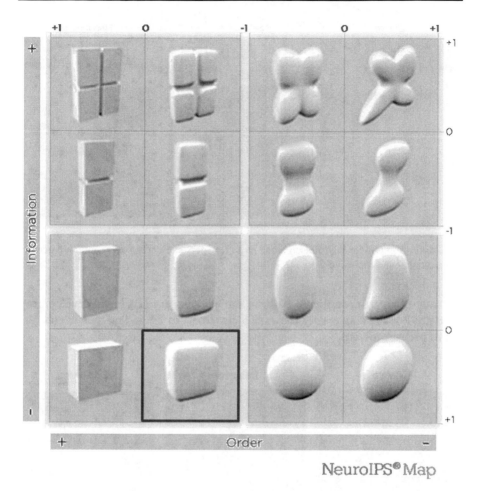

Fig. 7. Briefing tool for structuring communication

Short, succinct and straight to the point is the best way to provide Jahn Reisen customers with information, ideally supplemented by an abstract food for thought, as future trends and possibilities interest them greatly.

It is necessary to be prepared to discuss the pros and cons of the various options when talking to them. A bureaucratic approach scares them off.

With this description it is possible to optimize communications in many ways. To this end Figure 7 shows a briefing tool for developing communication.

The framed shape bottom left comes closest to a Jahn Reisen customer.

After completion of study results were used for fine tuning the catalogue's design, for example. An exclusive, cool appearing hotel ambience is paramount for the ENTP-type. Figure 8 shows the evolutionary brand development of Jahn Reisen based on this briefing.

Before	Relaunch	Fine-tuning

Fig. 8. Profiling of the Jahn Reisen brand to a clear personality (extrovert, intuitive perception, rational and exclusively oriented)

By way of comparison Figure 9 shows the evolutionary ITS brand development. The ESFP brand core mainly shows ITS customers to fall in line with the character of package tourists.

Before	Relaunch	Fine-tuning

Fig. 9. Profiling the ITS brand into a clear personality (extrovert, perceiving reality, primarily making decisions based on feelings, driven by flexibly)

5.3   Travel Agents Needs Compared to Those of Package Tourists

The study at travel agency level predominantly involved female participants (71.5%). This reflects the industry's typical gender structure. The age distribution corresponds, just like the package tourists, approximately to the general population.

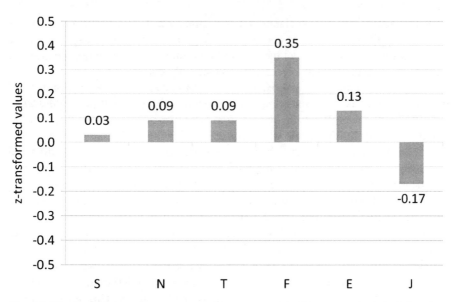

Fig. 10. The implicit travel agent personality systems (N=130) as standard deviation from the German norm (N=100,000)

Figure 10 shows the characteristics of implicit travel agent personality systems. The dimensions perceiving and extraversion are the same for travel agents and the group of package tourists. The dimension of perception "intuition" approximately corresponds to the value of package tourists, whereas the dimension of perception "sensing" (0.03) is clearly behind the value of package tourists by a delta of 0.20. Therefore travel agents are less interested in facts and details as well as a step by step procedure. The dimension of decision "feeling" clearly emerges with a standard deviation of 0.35. Travel agents tend to base their decision on instinct rather than proceeding from any analysis. They think they need to defend their values and are inclined to make decisions subjectively. For them interpersonal contacts, fostered by trust and empathy, are vital to them. So in essence travel agents can be regarded as ENFP type.

5.4   General Description and Motivation of Travel Agents as Well as Potential Discord Among Travel Agents

Travel agents are amicable, relaxed, warm-hearted, full of energy, authentic and have people skills. Travel agents generally have a relatively strong profile. To attune to many different customers and communicate with them is a special ability of travel agents. They are good at observing and forefeeling what is going on inside of someone, sensing feelings and giving words or actions a special meaning. In general problems which suddenly appear are dealt with their talent for improvisation at astonishing ease.

Travel agents are very sociable and sometimes can't wait to tell others about their extraordinary adventures and experiences. This can by all means take some time. Yet simple stories are not the focus of their tales but rather more the effort to pass on truths about human impressions and strongly convincing.

It might happen that they just fixate on own ideas and expectations whilst consulting package tourists. They sometimes think too much about what they would want.

Considering the profile it can be suspected that travel agents, despite their inspirational aura, can't fulfill the role which customers give them. Customers expect clear decision-making support when choosing their holiday destination. The structured and analytical thinking which is required isn't the travel agents" strong point and thus demotivated them. Travel agents prefer working associatively and emphatically. Travel agents have varied interests – they are also interested in themselves. Personal target setting and development have a high priority compared to other types.

Travel agents like to think about "nice things". Working daily with stories and pictures of holiday trips really meets the neuropsychological function, which have developed them most. From their point of view it is necessary to experience intensive emotions to have a fulfilling life. They find it great undergoing new and positive experiences. Trips and parties offer such new stimuli. This is why travel agents get bored with everyday life. They take great joy in expanding their practical expertise and applying it in direct communication when working together with other people. Every sales conversation is a small adventure, an impression and is different just like people. Wealth of experience which touches people's hearts interests travel agents. Sales conversations often have a pleasant atmosphere as people look forward to their holiday which is in line with the travel agents nature. Travel agents can only flourish in a good atmosphere.

5.5   Communication with Travel Agents

Experiencing hands-on is the best way of learning for travel agents. Things they experienced themselves sticks with them can be reproduced easily. However they find it difficult to acquire knowledge from catalogues. Therefore stories backed up by picture material are very important for communicating with travel agents. Written detailed information should be kept as short as possible.

They like energetic and inspiring communication and appreciate it when being acknowledged and recognised. Ideally information is passed on in short separated sequences. Markers can be applied to better emphasize single sequences. For example a model elephant could be put on the table during a sequence of Africa.

Training must be very lively, varied and unconventional, backed up by picture material, surprising actions and stories. Events not only demand the most expensive champagne because cheaper options may be more effective, tailored and come with a surprise element. The possibility to introduce oneself, ask questions and opportunities to further develop greatly improve an event program.

An information overload is the greatest mistake when communicating with travel agents.

6 Discussing the Results and Practical Implications

The subconscious personality profile of brand users tells a lot about needs, motives and resulting preferences which can be used together with visual briefings for structuring communication. Furthermore the mean profile of brand users tells something about the brand personality. These should be strengthened by consistent communication specifically for target groups because matching brand and customer profiles are an essential factor for achieving brand preference (Florack & Scarabis, 2007). By using empirical methods of unconscious personality profiles of brand users, differentiation of brands can be applied more specifically and effectively (Markgraf & Scheffler, 2011).

References

Florack A & Scarabis M (2007) Personalisierte Ansätze der Markenführung. In: Florack A, Scarabis M & Primosch E (eds), Psychologie der Markenführung. Munich: Vahlen

Markgraf I & Scheffler D (2011) Differenzierung von Marken durch Neuromarketing am Beispiel der Tourismusbranche. Working papers of the Nordakademie, 2011–2003, Internet publications by the Nordakademie: http://www.nordakademie.de/arbeitspapier +M55765093672.html

Mund JW (2006) Tourismus (Third Edition). Munich: Oldenbourg.

Nisbett RE & Wilson TD (1977) Telling more than we can know: Verbal reports on mental processes. Psychological Review, 84, 231–259

Scheffer D & Loerwald D (2008) Measuring personality characteristics with the Visual Questionnaire (ViQ) – Attraktivität als Nebengütekriterium. In: Sarges W & Scheffer D (eds), Innovative Ansätze für die Eignungsdiagnostik, (pp 51–63). Göttingen: Hogrefe

Zaltman G (2003) How customers think. Boston: Havard Business School Press

Product Development, Marketing, and Distribution

Kirsi Hyvärinen, Dusanka Pavicevic, and Dennis Hürten

1 Introduction

Change management is generally understood as a planned approach to transition-ing organizations from a current state to a desired future performance. In travel and tourism, the challenge is multiple, because the object of change is similar to a biotope: Give a cell favorable conditions to develop and grow, and it will do so, but the direction or scope might not be quite what you intended. Nevertheless, good niche products and services may by all means be born in biotopes where older forms of existence have not yet filled the space.[1]

Furthermore, the never-ending challenge in service-chain-dependent "biotopes" such as tourism is the fact that your customer wishes to buy his best time of the year, often pre-packaged, and judges the overall experience by its details:

- The first impression is decisive, from the questions at home of "What to do?" and "Where to go?" to the arrival experience, such as "Great. Our 10-day cy-cling trip started in a downpour." *You can change some things, but not all.*

- The holiday experience is never fully repeatable; thus the customer experi-ence might range from "This is why I came!" to "I wish I were ... (some-where else)," even though the destination and the suppliers are the same. *Pick the cherries your guests will talk about, whether big or small, but let them be tasty. Work with the ones who are willing and supportive of your strategy.*

- The last impression remains, and in the best case it sounds like: "I'd like to come back again." Or, if a link in the service chain was a bad one, its meaning will tend to become magnified, because the traveler will tell friends and relatives about it more often: "Nice country, but what horrible service on the flight back home ...!" *One bad apple can spoil the whole bunch.*

---

[1] Compare with Müller, Mokka: *Das vierte Feld – Die Bio-Logik der neuen Führungs-elite*, Munich 2001, p. 30ff.

To manage more mechanical travel service details, the psychologies of personal delivery and client expectations during the experience should be challenge enough. As a destination management organization, it surely took a great deal of work to get to where you are today. Why is it necessary to deliberately think about any changes at all? As Wayne Gretzky, one of the best ice hockey players of the 20th century put it: *"I skate to where the puck is going to be, not where it has been."* Consequently, it's no surprise that the number one reason for change is to keep pace with the marketplace. Every destination finds itself in a completely different environment than it did thirty or even ten years ago: fierce competition; sudden events, capable of reversing the global economy virtually overnight; global environmental factors; savvy clients with individual wishes; traditional information and booking channels increasingly replaced by direct communication, resulting in the need for destination management and marketing organizations to successfully interact in social networks. For many countries and regions, sheer dependency on spending will be the driver for change. In this context, linkages and opportunities with other sectors of the economy are far too often considered only superficially or neglected.

"According to Darwin's *Origin of Species*, it is not the most intellectual of the species that survives; it is not the strongest that survives; but the species that survives is the one that is best able to adapt and adjust to the changing environment in which it finds itself."[2] Talking about travel destinations, "strength" often translates into big product investments, and "intellectual" into multi-level marketing budgets. But what if you have an emerging, declining or in any other way "weak" destination to start with? Or you are experiencing severe setbacks due to external circumstances beyond your control? Or perhaps your key partners – buyers, sellers, politicians, developers – are not buying in to what you see as the keys to success? Or your clients are not the ones you actually wanted?

Don't let the frantic pace of operations replace strategic calm. Take the time to step back and appraise the situation; sharpen your sense of the existing challenges and new ones to come. Set priorities. In the following, some of the key areas to look into are introduced, based on practical experience and useful theoretical frameworks. In the last section, a case study provides insights into the experience gathered in an emerging European destination, Montenegro, in its process of change.

---

[2] Megginson, Leon C.: "Lessons from Europe for American Business," *Southwestern Social Science Quarterly* (1963), at: www.darwinproject.ac.uk.

2   Change Requires Leadership

The primary and fundamental challenge of destination leadership is to align all resources to serve the respective values, mission, vision and strategy.[3] In many cases, these need to be determined in the first place. Are the mission (why we exist), values (what we believe in), vision (what we want to be) and strategy (our name of the game) defined at all, and are they well communicated? Do your pivotal people and partners understand, support and live by them? How can you tell?[4]

Strategy should not stand alone as something only the management and board deal with. Strategy is one step in a logical ladder that moves an organization from high-level mission and vision statements to the individual level. It begins in the broadest sense with the mission of the organization. The mission must be translated so that the organizational operations and the activities of individuals are aligned and mutually supportive. The leadership, managers and the general working system should ensure that this translation is done effectively. "Success comes from having strategy become everyone's everyday job."[5]

Thus, change often begins within your own organization. In most cases, there is a history to deal with, and you may need to move your change cornerstones in order to respond to situations from the past and to future requirements at the same time. Areas in the destination change process may include, for example:

- Strategic changes (national policies, source markets, target groups, product range)
- Mission changes (mission, vision, positioning, brand)
- Changing attitudes and behaviors (education, staff, suppliers)
- Operational changes (including structural changes)
- Technological changes (aligning business and systems, introducing new marketing and sales channels)

The planning and decision-making process will affect success. Participative processes and decentralization take more time, but will help ensure sustainable implementation and growth. In this phase, it can be useful to involve a temporary leadership coach to help manage the change process. The scope of tasks for a coach and the results he or she should achieve can be linked with four main roles, depending on the mutual levels of knowledge and experience:

---

[3] Compare with Dym, Barry; Egmont, Susan; Watkins, Laura: *Managing Leadership Transition for Nonprofits: Passing the Torch to Sustain Organizational Excellence*, New Jersey 2011; p. 20ff.

[4] Compare with Kaplan, Robert S.; Norton, David P.: *The Strategy Focused Organization*, Boston 2001, p. 73.

[5] Ibid., p. 3.

- To feel out and localize the scope of change needs

- To act as a sparring partner or as a new approach developer, and in part also to be "the barking tree"

- To serve as a time-limited additional resource for the management and the organization

- To develop solutions[6]

Crucial for the leadership and management is the right mindset: challenges are opportunities. If you keep hearing: "Yes, I agree, but the problem is …" – full stop. What is *in reality* hindering you (or others) from moving from planning to action? Whatever the answer, stay customer focused.

In almost any destination, you will face a host of challenges that require attention, and, if urgent enough, they may lead to smaller change programs or to a larger transformation, depending to a great extent on the position in the destination life cycle.[7] The position often occurs in combination of two or more:

- Mature destinations (with little state involvement)

- Emerging destinations

- Declining destinations (in need of revitalization)

- Centrally planned economies

- Destinations with fragile environments and/or endangered species

- Countries with market perception problems

- Destinations with a dominant product

- Theme-focused destinations (adventure/culture …)

- Destinations with less favorable development opportunities

- Cities[8]

For nonprofit and public organizations it is just as relevant to express what the organization intends to do as what it decides *not* to do.[9] Otherwise, there will be "scope creeping," meaning spoken or unspoken (and often costly!) expectations of the organization by legislative, public or private partners, which will be impossible to meet without compromising the strategy.

---

[6] Compare with Huttunen, Pekka: *Onnistuneen konsulttihankkeen toteuttaminen*, Helsinki 2003, p. 19.

[7] Compare with Butler, R. W.: "The concept of a tourist area cycle of evolution: Implications for management of resources", in: *Canadian Geographer*, 24 1980: pp. 5-12.

[8] Compare with Tourism Development International: Presentation 10th Feb. 2011, ETC/UNWTO Handbook on Tourism Product Development.

[9] Compare with Kaplan/Norton, p. 133.

To ascertain the relevant opportunities, and to create successful and sustainable improvements in line with market requirements for innovation, differentiation and authenticity, you need to analyze the present situation, identify the opportunities, and prioritize your destination's overall economic goals and the specific tourism sector's objectives.

The basic management process of "analyze – plan – implement – control – learn" will stay the same. Don't skip any step, but it is still good to look for some shortcuts. They may result in better outcomes and/or in lower costs. Example: For basic tourism infrastructure, there is mostly no need to re-invent the wheel. Either you will come out with a square product or you are missing opportunities for savings and synergies.

For destinations, the path to sustainable competitive advantage will seldom lead to price leadership. A differentiation strategy is the most probable one. True differentiation arises from the choice of activities and the way they are performed all along the value chain.[10] With your stakeholders, a joint look through some "diagnostic windows" can help sort out basic work areas. For deeper understanding, better focus and meaningful results, add the simple question "Why?" after each statement made. You will come close to the *core purpose* of your organization. The core purpose must not be mixed up with the goals or strategies. They can and should change at reasonable intervals, but the "reason for being" will not.[11]

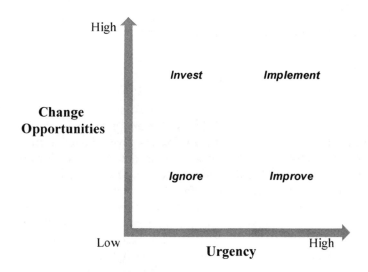

Fig. 1. Diagnostic window – change opportunities/urgency

---

[10] Compare with Porter, Michael E.: *Wettbewerbsvorteile: Spitzenleistungen erreichen und behaupten (Competitive Advantage)*, Frankfurt/New York 1992, p. 93ff.

[11] Compare with Collins, James C. and Porras, Jerry I.: "Building Your Company's Vision", in: *Harvard Business Review on Change*, Boston 1998, p. 30ff.

3   Change Starts with Results

The bottom-line outcomes must be the guiding star of any worthwhile destination change program.[12] With limited resources (financial, human, time) you need to focus on those initiatives that will produce the strongest impact on achieving the results within a defined time frame. Change must be made measurable. Instead of listing potentials, initiatives, programs and activities, a well-articulated strategy talks about the outcomes the destination is planning to achieve. If the change is made quantifiable and the activities of the organization are clearly derived from the strategy, the results in focus are the best legitimization for further funding. Therefore, the logic of a "result chain" or a "result wheel" must be agreed on right at the beginning:

- "To satisfy our mission, customers and financial partners, what business processes must we excel at?"
- "Why?"
- "To achieve our vision, what must our people learn, how should they communicate and work together?"
- "Why is it important?"
- "What's in it for them?"
- "If we succeed, how will we look to our financial partners?"
- "When?"
- "Which activities can make it or break it?"
- "How can we tell?" [13]

The change process goes through a series of phases. Skipping steps only creates the illusion of speed but never leads to meaningful results. Even though a great enough sense of urgency is important, each step will take its time. Serious mistakes in any of the phases – for example too little communication of the decisions and plans – can negate hard-won gains in the phases before.[14] The most important underlying support activity through all the phases is communication with internal and external customers, employees, business partners and the public. It is a fact that is often forgotten, but communication always goes both ways. Listening to your existing and potential guests, you will gain priceless inputs and intelligence that will help you to serve them better and to fulfill the core purpose of the organization.

---

[12] Compare with Schaffer, Robert H. and Thomson, Harvey A.: "Successful Change Programs Begin with Results," in: *Harvard Business Review on Change*, Boston 1998, p. 189ff.

[13] Compare with Kaplan/Norton, p. 133ff.

[14] Compare with Kotter, John P.: "Leading Change: Why Transformation Efforts Fail", in: *Harvard Business Review on Change*, Boston 1998, p. 3ff.

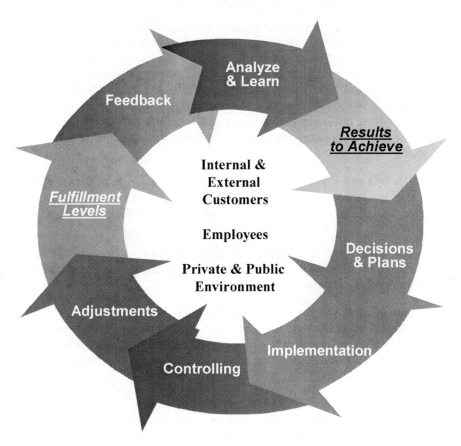

Fig. 2. Results in change process

3.1 Customer and Cost Perspectives

Knowing the values, needs and wishes of existing and future clients is the most powerful basis for all future marketing activities. To fully benefit from market intelligence, you need to comprehend the complexity of the strategic marketing approach.

For destination management organizations this usually means that they have to grasp the idea that tourism product development and putting these products on the market are not two separate things.

After the transition from a sellers" to a buyers" market, market intelligence has become an essential precondition for profiling the challenges of private and public tourism organizations. And even though obvious, the utilization of market intelligence is often missing completely, or its use is inadequate. The decision to approach the market strategically will entail a lot of new tasks. The most important one is to understand that the relevant market has to be analyzed continually. Consequently, the collection and evaluation of (primary) market data can be attached to the need for a budget for market research within the enterprise or organization. Such a budget is designated for two tasks. The first one is to finance the data that needs to be collected. The second one is to finance the market research workforce within the organization. Either the existing workforce needs to be trained in tourism market research – and they should be nothing less than the internal "high potentials" – or tourism market research expertise must be hired or outsourced.

In the strategic marketing process you need to analyze the current market position of the enterprise or organization in the relevant market(s) and in the industry as a whole:

- The first step is to analyze the segment of the market in which the destination has its core business. The focus of this analysis is usually on regional or thematic aspects of the tourism market, such as activity or cultural tourism.

- The second step is to analyze the global (tourism) market with the economic, social, ecological, technological, and political environmental factors for the destination in mind.

The results of these two deliver valuable input for a SWOT analysis. A SWOT analysis should never represent only the views of the staff involved in the creation process. Specific market data with views of existing and future guests needs to be the starting point: Guest satisfaction surveys or market potential surveys can deliver the needed insights.

- Finally, a sincere analysis of Porter's five competitive forces and their meaning for a destination will help shape a strong strategy and illuminate which results must be prioritized for best possible cost effectiveness (threat of established rivals; threat of new entrants; threat of substitute products; the bargaining power of suppliers; and the bargaining power of customers).[15]

Depending on the type of destination (with focus on activities, culture, nature, etc.), different types of knowledge are important for the marketing process. The same is true for the relative size of a destination, defined by the proportion of reinvestment in tourism and the economic dependency on tourism (as a percentage of GDP). The position in one of the fields should have direct impact on the budgeting of product development and marketing. Decision-making becomes risky if your destination is in a growth phase and/or highly dependent on tourism and you do not invest enough in knowledge transfer.

It costs far less to create a good knowledge and qualification basis than to pay for the consequences of wrong decisions, which may have been made for political or other reasons. These often do not serve the customer perspective. It takes some stamina to bear the uncertainty period between investment and revenue, which

---

[15] Compare with Porter, p. 22ff.

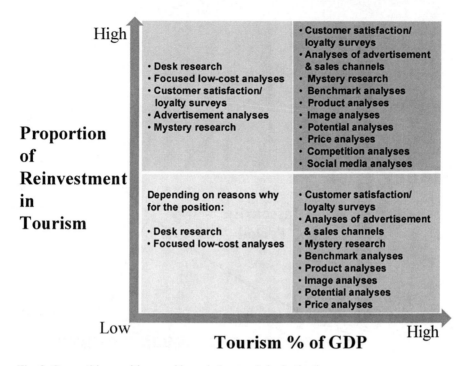

Fig. 3. Competitive positions and knowledge needs in destinations

occurs between the beginning of any change program (analysis, planning) and the later phase when results start becoming tangible at the micro- and macroeconomic levels.

Therefore, operational efficiency and the right choices for spending are crucial for legitimizing support. Important "customers" for a destination management organization are the ones that provide the funding: in the first place guests, but also private businesses, taxpayers and public funds. On the one hand, the organization must strive to meet the objectives of those sources as well – ultimately citizens.[16] On the other hand, the costs-by-cause principle must apply to the distribution of tax incomes from tourism. Successful organizations are all too often unfairly punished by budget cuts while tourism tax incomes are used for filling gaps in other areas.

3.2   Contribution and Cooperation

A "mechanical" perspective on change management focuses on observable, measurable business elements that can be improved, including business strategy, processes, systems, organizational structures, job roles etc. However, more than

---

[16] Compare with Kaplan/Norton, p. 133ff.

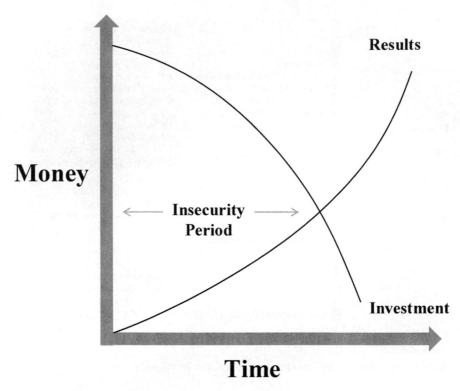

Fig. 4. From investment to results

Source: Own figure, based on Tourism Development International: Presentation 10th Feb 2011, ETC/UNWTO Handbook on Tourism Product Development

anything else, tourism is a people business. Therefore, the ability to manage the people side of change in the presence of a new culture and new values is a major contributor to successful change.[17] In the destination context this means that first of all internal, organizational and individual change must be managed. Second, horizontal and vertical cooperation opportunities must be actively sought to nurture sustainable and efficient growth. Third, external networking is essential to enable thinking out of the box. With a variety of options available, you can better react to unforeseen developments.

Contribution and cooperation are the essence of any synergy – by definition the interaction of two or more elements: the combined effect is greater than the sum of the individual effects.[18] Like no other industry, tourism spurs the development of other sectors, such as rural, building and construction, retail, real estate and

---

[17] www.change-management.com.

[18] The term *synergy* comes from the Greek (*syn-ergos*, meaning "working together"), http://de.wiktionary.org.

infrastructure, healthcare and personal services, credit and insurance industries, design and manufacturing, as well as information and communication technology (ICT). Vice versa, intelligent cross-sector cooperations will foster the development of tourism, and can add value to the guest experience. For example, a cultural sightseeing walk can be supported with mobile applications; a unique, authentic experience is multipled by integrating local produce, people, habits and storytelling into the basic tourism product of travel and accommodation. "Modern tourism is inspired by people's desire to commune with their fellow humans, celebrate each other's cultures, and take delight in the wonders of nature that surround them. Ask anyone why they travel, and this is what they most likely will say."[19]

There is no sharing or good cooperation without good communication. With today's technical possibilities, the rule of thumb could be: Always communicate in such a way that others can participate. And: "Always act so as to increase the number of choices."[20] Involve internal and external specialists in round tables, young and old, big and small businesses; turn outsiders into insiders and make your staff empathize with clients; support strong, open networks. Involve different styles, cultures and thinking, and foster views unlike your own.[21]

"*I am who I am through you.*" These words are from the owner and founder of a small African adventure travel business, today very successful, quoting the ancient African wisdom of *Ubuntu*, which states that the very essence of being human is the fact that we cannot exist in isolation. Sharing knowledge, experience, market intelligence, product ideas and even financials and resources is often perceived as a threat because of its assumed effect on one's profitability and competitiveness. In today's thinking, however, contributing and sharing are sure to stimulate ideas, to inspire, and to create "power in numbers." Generosity of spirit reveals potential, and can lead to lasting bonds between yourself and your peers. Passing along knowledge and vulnerabilities, and letting people grow with the benefit of your expertise or resources, also help you and others to grow, both within and outside your sphere of influence.[22]

"What I do affects the world. When I do well it spreads out. Sharing is at the very heart of our industry: Each day we share what we love with our clients and our guests. That is easy. So why don't we share with each other? Broadly speaking we are chronically sharing-averse in the travel industry. Do we ... reveal our financials? Disclose our mistakes? Share our customers? Offer the competition a helping hand? Offer to joint market with our competitors? Partner with our suppli-

---

[19] Rifkin, Jeremy: *The Empathic Civilization*, Cambridge 2009, p. 441.

[20] "Der ethische Imperativ: Handle stehts so, dass die Anzahl von Möglichkeiten wächst" von Förster, Heinz: *Wissen und Gewissen: Versuch einer Brücke*, published by Siegfried J. Schmidt, Frankfurt Main 1993, p. 49.

[21] Compare Hopkins, Rob: *The Transition Handbook*, Totnes 2008, p. 147ff. and Müller, M.: p. 360.

[22] Ms. Nicky Fitzgerald, a luxury adventure travel industry and marketing veteran in her speech at the Adventure Travel World Summit 2010 in Aviemore, Scotland.

ers? Partner with our communities? Are we open and available to others? Affirming of others? Interconnected with all stakeholders in our industry? Generous? If we contribute, share, work together, learn from each other and pay it forward, we will be successful and through that make a difference – at the end of the day, a sustainable difference to the planet."[23]

Strong competition is good for you, for it provides a chance for qualitative growth. And the more "joint noise" your competitors make, the more your services will get attention, and vice versa. However, we tend to see the *success* of competitors as a role model and to copy and learn from them. But it is even better to look for the failures and to learn from the *mistakes* of others, to anticipate pitfalls and to share your own mistakes as well. This will save time, energy and money for all involved. It is better to risk small errors on the way, in order to gain experience and protect yourself from the truly dramatic big mistakes. "The man who makes no mistakes does not usually make anything."[24]

Destination managers need to develop a "cooperative IQ," i.e. to learn how to cooperate; despite our innate tendencies, to learn to cooperate in the face of competition, to look outside the travel and tourism industry and current circles of navigation. Innovation and creativity, coupled with increased interest in contributing and sharing, can lead to powerful and fruitful alliances with often unlikely individuals, businesses and organizations.[25]

You can transfer experience from one area (which you have been comfortable with) to new ones (leaving your "comfort zone"). As an example, in the Black Forest, farmers" wives, hoteliers and tourism service providers have always been good at sharing – at the local level, among neighbors and friends, each in their own "zone." The idea of shared meetings was however new. After some time though, the "quality circle on family holidays" started to deliver innovative service ideas to destination management. As a contributor of "content" to the big picture (the vision of the regional tourism organization), these service ideas helped propel a whole region to the status of one of the most successful destinations for family holidays in Germany. [26]

For the globalized world, international and cross-border cooperation is becoming increasingly meaningful. One good example is the Adventure Travel Trade Association. Focused cooperation and overcoming fragmentation can improve the situation of all businesses, and especially small ones. This has been the experience at the yearly Adventure Travel World Summits, with the Brazilian Association of Adventure Tourism Companies (ABETA) as one of the hosts in 2008, Canada's Québec province hosting in 2009, or the 31-day packages prepared for practical

---

[23] Ibid.

[24] Schaffer/Thomson, p. 181.

[25] Mr. John Sterling, Executive Director, The Conservation Alliance, in his speech at the Adventure Travel World Summit 2010 in Aviemore, Scotland.

[26] www.familien-ferien.de/Regionen-Orte/Schwarzwald/Ferienland-im-Schwarzwald.

familiarization on the "Day of Adventures" in and around Aviemore, Scotland, in 2010. All this has been made possible not by huge funds, but mainly by seamless cooperation between all profit and non-profit partners.[27]

In post-conflict areas, cross-border cooperation can help enhance the quality of life. Jointly developing the touristic attractiveness of a region enables the economic revitalization and social cohesion of local communities. "Peaks of the Balkans" in the border area of Albania, Kosovo and Montenegro is one such project; here, guests will soon have the chance to discover natural and cultural treasures of the Prokletije mountain region and the truly unbounded hospitality of its people.

In this same spirit, World Tourism Day was celebrated in 2011 under the theme *Tourism – Linking Cultures*, which is intended to highlight tourism's role in bringing the cultures of the world together and promoting global understanding through travel: "This interaction between people of different backgrounds and ways of life represents an enormous opportunity to advance tolerance, respect and mutual understanding" (UNWTO Secretary-General, Dr. Taleb Rifai).[28]

3.3   Credibility and Consistency

Simple, but in its true meaning often forgotten: You need a product to market and sell. A destination brand is in the first place a huge promise of that product. It gives guests security that their dreams, wishes and needs will be met. To be successful, the destination must continuously keep the core essence of this promise. To stay fresh, the brand must be filled with credible content, i.e. with the right kind of "hardware" (infrastructure, accommodations) and "software" (innovations, information, services). Guests must be able to experience the brand with all their senses, above all through compatible, strong ingredients and in adequate events, which transfer the same message over a longer period of time.[29]

Most importantly, the promise of the brand and its content must match the values and contexts of the target groups. This might seem like a bigger challenge for developed or declining destinations, which may need to think about more complex turnarounds. In emerging destinations, however, the pressure for fast developments may lead to a loss of focus, wrong decision making, uncontrolled actions, or a "rain dance" of activities, without truly understanding and delivering to the target groups. Soon, a downward spiral of not keeping promises to either the customers (of the core values and experiences promised) or to the suppliers (of growing and better-distributed incomes) would set in.

---

[27] www.adventuretravel.biz.

[28] http://media.unwto.org/en/press-release/2011-06-20/tourism-linking-cultures-unwto-launches-world-tourism-day-2011.

[29] Compare with the first results of Destination Camp 2011: "Kann Oberstaufen Coca Cola sein?," at: http://blog.kmto.de/veranstaltungen/die-marke-im-tourismus-ergebnisse-des-destinationcamp-2011.

An early and, upon closer scrutiny, very interesting example of the need for change and the efforts to effect it has been demonstrated by the island of Mallorca. After getting off to a successful start (1960s to 70s) in Spanish charter tourism, the island very quickly developed into the number one holiday destination for Germany and the UK. Due to what was already at that time a progressive infrastructure as well as optimal established flight connections, Mallorca was adopted by German tourists virtually as "their" island, and for tour operators as their cash cow, the cornerstone of their business. At the beginning of the 1980s, however, the figures started to drop drastically. One major reason: due to mass tourism and corresponding reports in the media, the image of Mallorca had turned into that of an "island for cleaning ladies" (in German: "die Putzfraueninsel"). New values and more diversified target groups had developed over the years, somewhat unnoticed by the destination. Soon, it was not chic anymore to tell your friends that you spent your holidays in Mallorca. The many "fans" of the island (the mass tourists) mostly spent their time in huge (not really luxury) hotels at the beaches and didn't explore the full beauty of the island. A new generation and target groups had to be pleased – "the other Mallorca" became the new message. Today, some three decades later, Mallorca's touristic product covers a huge variety of needs, well managed and catering to diversified expectations. This transformation has not taken place only on the surface, but also in the depths of the core ingredients of a successful destination. And the advice for flourishing through change continues to be: Never rest on your laurels![30]

The most delicate step is from the promise to sales, from sales to delivery. What does a destination brand sell? An emotional promise, and the security of a specific bundle of feelings and experiences. Your brand must be tangible for the customers you want to reach. The strength of a brand depends on the relevant target groups. The differences between successful destination brands and those of consumer goods such as Coca Cola are numerous, and can all be seen as opportunities:

- Consumer goods must be mass-compatible, but not destinations
- Coca Cola: unchanged recipe versus a changing choice of heterogeneous services
- Consumer goods: the control of the brand can be strong; in destinations, brand management needs supporters (see previous section)
- For destinations, internal communication (accommodation and service providers, politics) is much more important, more difficult, and never-ending
- Source markets and travel seasons influence brand management
- Consumer goods must invent a unique selling proposition (USP); destinations may have many of those, but must focus and spotlight

---

[30] Interview with Ms. Anita Meier, Founder and Director of the Mallorca Information Office in Frankfurt 1982-89, consulting, destination management, marketing and PR, later also responsible for the Balearics.

- Destinations need to provide a lot of information, although its use is on a short (in worst case one-time) basis

- Consumer goods do not depend on cooperation, destinations do[31]

Most importantly, you cannot "touch and feel" a destination unless you are there. With social networking, a destination can change this, before and after travel. This can lead to higher emotional involvement and strengthening of the brand, despite less frequent use than of the daily preferred consumer goods. Emotional involvement is the strongest challenge and the best opportunity for a destination: very few other industries can truly sell strong feelings.[32] Furthermore, the closer the destination, the more people know about it. The further away the clients are, the more cliché-based the image. You can overcome this by cultivating involvement: different source markets can be reached with consistent media work.[33]

The responsibility for brand management must be clear though, in order to enable synergies and better visibility of a whole region across sectors, not only in tourism. Interesting European examples of integration are the Nature Park Black Forest Central/North[34] and South Tyrol.[35] Both demonstrate the ability to work with authenticity, to "stage manage" the brands without compromising the core promise (for which there are plenty of other examples in the world). This kind of integration allows enough freedom for sub-brands to deliver their umbrella-brand-compatible products and services with unique local content. The small town of Rakvere in Estonia has started to translate its traditions of "power" and "music" into contemporary interpretations, such as the first Estonian Punk Song Festival. It's not quite a champion in tourism brand management yet, but the ingredients are there, including supportive local politics.[36]

Knowing your target group(s) is one of the most important prerequisites for longstanding marketing success. Once you have decided to use target group marketing, a lot of new questions arise at first for destination management organizations, for instance:

- Do our targeted customers know our destination? Do they know our tourism offer?

- What do they think of our destination? Would they recommend it to other guests?

- Are the people we defined to be our target group(s) and the people who are interested in our products actually the same?

---

[31] Destination Camp 2011 results.

[32] The "Citysherpa" concept in Helsinki made use of "face-to-face" contacts well before the Facebook & co. boom: www.guardian.co.uk/travel/2007/jun/30/saturday.helsinki.

[33] Destination Camp 2011 results.

[34] http://en.naturparkschwarzwald.de/regional.

[35] www.suedtirol.info.

[36] www.visitestonia.com/en/holiday-destinations/city-guides/rakvere and www.rakvere.ee.

Important marketing tools for answering these kinds of questions are *customer typologies*. Since the tourism offer exceeds the tourism demand, they play an important role in tourism market segmentation. Common methods for segmenting the tourism market into more or less homogenous submarkets can be divided into socio-demographic, geographic, psychographic, behavior-related and psychographic approaches.[37]

Therefore, it is all the more surprising that, even with the plenitude of different segmentation approaches existing, there are few market segmentations available that divide the tourism market into valid segments. As a result, cross-sector typologies are often used to fill this gap. The problem here is that, for the use within certain industries, general typologies often tend to be too unspecific. Therefore, adequate state-of-the-art tourism typologies need to be developed.

One such up-to-date tourism typology using psychographic tourism market segmentation was developed by the research and consulting firm Trendscope from Cologne, Germany. Unlike existing socio-demographic, thematic or geographical segmentation approaches, the so called *Trendscope Traveller Types* provide comprehensive information about the relevant motives, desires and values of the travelers. In the creation of this typology the entire travel process from the initial holiday wish to the information and booking process, the journey itself and the individual follow-up of the journey was considered. Thus, all kinds of tourism destinations receive valuable information on the psychology behind information processes and accounting decisions and are therefore enabled to focus on the needs and values behind customers" booking decisions, allowing an accurate optimization of product development as well as marketing and sales activities.

With the Trendscope Traveller Types, the following questions for example can be answered:

- Target groups, e.g.: How are tourism target groups characterized? Where to find the appropriate target groups and how to win new customers?

- Product, e.g.: Do the current products meet the customers" wishes? How exactly should they be developed in the future? How to achieve better customer loyalty?

- Pricing, e.g.: What prices can be achieved within different target groups?

- Communication, e.g.: Do the corporate identity and corporate design fit the targeted customers? Through which channels can the target groups be reached?

- Sales, e.g.: What are the preferred booking channels of the target groups?

---

[37] Compare with Rück, Hans and Mende, Marcus: "Innovations in Market Segmentation and Customer Data Analysis," in: Conrady, Roland and Buck, Martin (eds.): *Trends and Issues in Global Tourism*. Berlin/Heidelberg 2008: pp. 137-148.

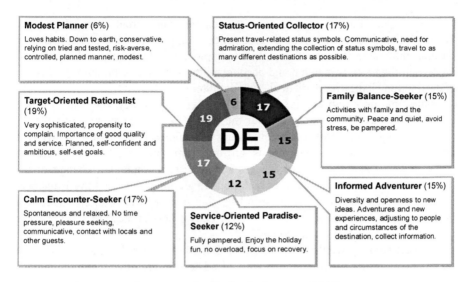

Modest Planner (6%)

Loves habits. Down to earth, conservative, relying on tried and tested, risk-averse, controlled, planned manner, modest.

Status-Oriented Collector (17%)

Present travel-related status symbols. Communicative, need for admiration, extending the collection of status symbols, travel to as many different destinations as possible.

Target-Oriented Rationalist (19%)

Very sophisticated, propensity to complain. Importance of good quality and service. Planned, self-confident and ambitious, self-set goals.

Family Balance-Seeker (15%)

Activities with family and the community. Peace and quiet, avoid stress, be pampered.

Calm Encounter-Seeker (17%)

Spontaneous and relaxed. No time pressure, pleasure seeking, communicative, contact with locals and other guests.

Service-Oriented Paradise-Seeker (12%)

Fully pampered. Enjoy the holiday fun, no overload, focus on recovery.

Informed Adventurer (15%)

Diversity and openness to new ideas. Adventures and new experiences, adjusting to people and circumstances of the destination, collect information.

Fig. 5. Trendscope traveller types – distribution in Germany (2010)

The Trendscope Traveller Types are based on a basic research project that analyzed and systematized the entire travel process – from preparation to realization and follow-up – and the psychology behind it. The result is a comprehensive target group approach, which enables tourism destinations to focus their market activities on the motives, wishes and values of their relevant target groups.[38]

Similarly on the offer side: dividing a destination into development clusters is an effective way of promoting the strengths of each region and spreading the socio-economic benefits of tourism. This approach must however be used in a highly integrative manner, to avoid internal "protection" and barriers to entry (the opposite of sharing and cooperation).

Consistency in such new approaches calls for empowerment (make the right decision for the customer), accountability (take ownership and pride in your work), and continuous improvement (look for ways to improve everything you do, everyday). A new culture can evolve in many of today's destination management organizations, where a new generation of leadership and employees are taking ownership and responsibility for their work, have pride in their workmanship, look to improve their work processes, and feel empowered to make decisions that improve the bottom-line results, in keeping with the core purpose and promises made.

---

[38] *Methodology:* For the development of this psychographic market segmentation, two successive methodological steps were taken. In the first step, psychologically trained interviewers conducted more than 40 in-depth interviews and explored travel motives, travel interests and preferences, travel behavior etc. in the different phases of the travel process. The interviews had an average length of two hours. Based on this pre-study, 28 statements were developed, which were later used for the assignment of the respondents to the seven Trendscope Traveller Types.

So why does this sometimes sound so remote from actual daily experience? The evolution from the traditional values of control, predictability and consistency – values that made change rather simple to implement – to the new values focused on decentralization, ownership and accountability, have made the implementation of destination change more difficult, and definitely worth doing.

Today, partners and employees question their day-to-day activities and are hopefully rewarded for doing so. But the way to resisting change initiatives is then just around the corner. The new values of business today require a different approach to change: the response is shifting from "Yes, sir" to "Why are we doing that?" – and the change leaders should listen. Managing change in the "old" value structures involved only announcing change and making it happen. But if a destination change leadership tries this same approach today, there will be shouting and gossiping "Why?" – "How does it impact me?" or: "If the horse is dead, why are we trying to ride it?"

The best strategy would be to dismount. But where bad politics are paired with short-term interests, other strategies with "dead horses" are often tried:

1. Buying a stronger whip.

2. Appointing a committee to study the horse.

3. Threatening the horse with termination.

4. Changing riders.

5. Arranging to visit other sites to see how they ride dead horses.

6. Lowering the standards so that dead horses can be included.

7. Reclassifying the dead horse as "living-impaired."

8. Finding excuses why some live horses should be kept out of sight.

9. Hiring outside contractors to ride the dead horse.

10. Harnessing several dead horses together to increase speed.

11. Providing additional training to increase the dead horse's performance.

12. Doing a feasibility study to see if lighter riders would improve the dead horse's performance.

13. Declaring that the dead horse carries lower overhead and therefore contributes more to the bottom line than some other horses.

14. Rewriting the expected performance requirements for all horses.

15. Promoting the dead horse to a high position.[39]

The conclusion that might be derived from the above for credible and consistent change efforts is that it is good to delegate consulting and decision-making com-

---

[39] Compare with www.bpic.co.uk/articles/deadhorse.htm.

petency, with responsibilities for the outcomes, to the lowest level at which a sound judgment can be made – and reward the results. At the end of the day, "it is the front-line employees that ultimately execute the new day-to-day activities and make the new processes and systems come to life in the business."[40] In destinations, some "top down" navigation is good, but sustaining credibility and consistency requires decentralized capability growth and empowerment.

4   Case Study Montenegro

The following experience describes one of the world's youngest states, Montenegro, in its process of change from a forgotten travel destination in the post-conflict region of the Balkans with some collective perception problems, to today's path toward an emerging, decentralizing and theme-focused setting. The country is on the one hand endowed with some of the finest natural and cultural prerequisites for a distinctive travel destination; on the other hand, it has been and remains under strong pressure to transform socially, economically and ecologically. After demonstrating a case for a destination with a sound claim to be strategy-focused, the second part of the case study continues with the presentation of a tangible success story in sustainable product development and marketing.

4.1   Conditions and Change Characteristics, 1990–2010

Back in 1990, some 11 million overnights in Montenegro were counted and 35% of the total number of guests came from abroad. After three wars in the region, the embargo imposed on the Federal Republic of Yugoslavia and the hyperinflation in Montenegro, objective reasons had contributed to a devastating downward spiral.

Montenegro had to face extremely unfavorable start-up conditions in 2000. The country's name was totally forgotten in all European markets. Its tourism product had been conceived by a socialist bureaucracy. It consisted of three major components: some anonymous but huge hotel blocks of low standard, an almost exclusive concentration on summer bathing holidays at the coast, and a former focus on mass tourism in the 1960s, 70s and 90s. Best known as a cheap, low-budget vacation spot, the image not only of Montenegro, but of the whole of former Yugoslavia's coast was not competitive. Its only *raison d'être* had been to earn foreign currency, and there had been very little concern about the social, cultural or ecological effects that might result from this type of mass tourism.[41]

---

[40]   www.change-management.com.

[41]   Interview with Mr. Johann Friedrich Engel, Managing Director Creatop GmbH, Germany, Advisor to Masterplan 2001/Update 2008, and to the Tourism Development Strategy to 2020.

For years, no foreign tourist crossed the border. Instead, the hotels had to host the many fugitives without receiving the necessary means for maintenance or upgrading. In the year 2000, almost nothing in Montenegro was in a position to compete with any other destination and to satisfy foreign guests. There were only 3,000–5,000 beds to offer to guests from western and northern Europe.[42] As a result, in 2001 only 0.5 million guests were registered, and stayed for some 4 million overnights. Only every fifth guest was a foreigner.[43] About 50% of all guests came and went unregistered. "Our plan is to prolong the season and to change the structure of tourists. In 10 years we plan to have 50% and in 20 years to have at least 75% foreign tourists."[44]

Since then, travel and tourism have played a central role in Montenegro's dramatic growth and transformation. In 2009, the estimated share of tourism in *direct gross domestic product* in total GDP of Montenegro was 10.0%. This share places Montenegro among the countries in Europe with the *highest direct contribution of tourism to the national economy*.[45] And the value is expected to continue to increase in the years to come. In 2010, over 1.2 million guest arrivals and nearly 8 million registered overnights were achieved – repeatedly preceded by pole positions in European and world tourism growth statistics.

The guest structure is changing. Today, three out of five overnights (61%) are by foreigners in the narrower sense.[46] Nearly every third foreign guest comes from the European Union (28%); among those, the German market has reached a double-digit growth rate (38%). A total of 14% of the foreign guests come from Russia. The former eight-week peak seasonality has been reduced remarkably – the shoulder seasons from March to June and September to November have grown stronger. A capacity of over 15,000 beds in the categories 3, 4 and 5 stars are available, meaning 41% of the total bed capacity, as of May 2011.[47] World-class foreign direct investment at the coast, for example Aman Resorts, Orascom Holding and Porto Montenegro, are flagships that any destination would be more than happy to integrate.

---

[42] Interview with Mr. Predrag Nenezic, Minister of Tourism, 24th July 2001, in *World Investment News*.

[43] Statistics made available by the National Tourism Organisation of Montenegro.

[44] Interview with Mr. Predrag Nenezic, Minister of Tourism, 24th July 2001, in *World Investment News*.

[45] Pilot Compilation of Tourism Satellite Account in Montenegro for 2009, Draft Report July 2011, p. 6.

[46] Serbian guests are counted here as "domestic" in order to provide comparability with the figures in 2000, when the country was still "Serbia and Montenegro." Otherwise, the figure would be 88%.

[47] Statistics made available by the Ministry of Sustainable Development and Tourism and the National Tourism Organisation of Montenegro.

The ingredients of Montenegro's transformation as a destination have been comprehensive, and the resulting change is profound in many fields. The most significant ingredients, which are all related to tourism, are the following:

- Core purpose, strategy, and vision (consistent, agreed on)
- Legal framework (tourism, investment, banking, taxation etc.)
- Investment policies (open, supportive)
- Infrastructure (electricity, airports, roads, railway, water, wastewater, solid waste management, ICT)
- Destination management organizational setup (reorganization, decentralization)
- Private sector (SMEs, NGOs, associations, clubs)
- Human capacities (education, exchange, foreign support)
- Public-private partnering (growing nationally and internationally)
- National and international cooperation and networking (more, and specialized)
- Product development (focused diversification)
- Branding and promotion (supporting new methods and channels)
- Market intelligence and monitoring (on the way "from zero to hero")

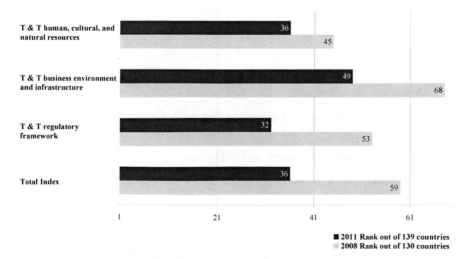

Fig. 6. Ranking in the Travel & Tourism Competitiveness Index 2008/2011

Source: World Economic Forum, Travel & Tourism Competitiveness Index, 2008 and 2011.

As one tourism development policy cornerstone, Montenegro's *Tourism Development Strategy to 2020*, adopted by the government in 2008 – with its "hard currency" objectives of diversification and quality growth – serves not only today's citizens, but also their children and generations beyond. Thus, the strategy and the manner in which it is implemented are perceived as critical for the nation's economic and social health. It is committed to assuring that tourism continues to be an engine for growth and that tourism catalyzes and generates the right kind of growth – smart growth, sustainable growth, balanced growth. The benefits of tourism are to not only be maximized, but also broadly distributed socially and geographically.

In today's turbulent times, it would be folly to attribute any certainty to forecasts made for a decade to come and beyond. Circumstances will change, and adjustments will need to be made, strategic sails re-trimmed. Committed as the country is to what has been identified as an appropriate, comprehensive and innovative strategic direction, as reflected in the documents and policies, there is equal commitment to maintaining a flexible approach that includes regular re-evaluation based on changing circumstances and emerging trends. [48]

Over the past decade, tourism has become Montenegro's most important industry. And today the country can be described as a "place where only few people have been, yet nine out of ten visitors want to come back."[49] Already at the end of 2009, the *National Geographic Traveler* identified the country as one of "50 Places of a Lifetime." In 2010, Montenegro was chosen as one of the three finalists for the World Travel and Tourism Council's *Tourism for Tomorrow Award* in the *Destination Stewardship* category.[50]

Nevertheless, some typical issues faced by any emerging destination continue to apply:

- Being in the growth phase of the destination life cycle
- Facing issues of enlargement: source markets, products, price ranges, distribution, communication
- Service levels not always corresponding with the promise of new upscale products
- Assisted through government and donor support but
- Vulnerable due to limitations both on staff and financial resources and
- Varying levels of implementation and enforcement of policies and laws

---

[48] Compare with Montenegro Tourism Development Strategy to 2020.

[49] Press release of the National Tourism Organization, March 2011.

[50] www.tourismfortomorrow.com/Winners/Previous_Winners_and_Finalists/Destination_ Stewardship_Award_/country-of-montenegro/index.php.

- Experience and lifestyle/value/perception gaps between own population and target groups. Example: the level of environmental awareness. While the desired visitors are already in the stage of sustainability as part of their lifestyles, most people in the destination are in the pre-contemplation phase.[51]

- Concerns about environmental and socio-economic impacts. Only with strong support by the public sector can the required behavioral change take place in the private sector.[52]

Particular attention is therefore paid to new product development and to "keeping the promise" of the brand *"Wild Beauty."* A homogenous product would appeal more to a geographically and demographically homogenous market. And customer homogeneity leaves a destination vulnerable to an accelerated destination lifecycle, in which today's hot destination can quickly become yesterday's place to go, seen to offer little of interest. As a starting point, Montenegro's compact diversity is truly unique. Breathtaking sceneries, rich history, three religions, three climate zones, rich wildlife, sandy beaches, glacial lakes and rivers, wild mountains and canyons, all wrapped up in only about 14,000 square kilometers. Not surprisingly, product and market diversity represents one pillar of the approach. A second, even more central, pillar is sustainability. It is inextricably intertwined with all aspects of the strategy and action derived from the ecological, social, cultural and economic dimensions.[53]

While Montenegro has consistently shown one of the fastest rates of tourism growth in the world, the challenge is not just to continue this growth, but also to assure that it is sustainable, balanced and brings both immediate and long-term benefits to the people of Montenegro, while protecting and preserving the natural assets that are the engine of tourism growth in the first place. Specifically, the vision is for Montenegro to become a differentiated, year-round Mediterranean destination – with a spectrum of unique attributes appealing to several key segments and niches of the mid- to upscale market. The efforts are focused on supporting an industry that is vibrant, multiseasonal and regionally balanced, supported by a strong and differentiated brand identity and sufficient air and accommodation capacity, not only at the coast, but in the central and northern regions of the nation as well.[54]

---

[51] Compare with Hopkins R., p. 84ff.

[52] Compare with World Economic Forum: *Travel & Tourism Competitiveness Index 2008*, "The Case of Montenegro: Steps toward environmentally sustainable tourism development," p. 32ff.

[53] Compare with Montenegro Tourism Development Strategy to 2020.

[54] Ibid.

4.2   Destination Transformation Supporting Sustainable Rural Development

In July 2007, the Montenegro Ministry of Tourism and the National Tourism Organization (NTO) launched an ambitious program called *Montenegro Wilderness Hiking & Biking*. At the time, Montenegro was catching the world's attention with stunning beaches, and the sun, sea, sand and sophistication of its coast. In contrast, its hinterland and mountainous north were barely known except to the most adventurous, with some rafting and alpine skiing. Furthermore, in the north, old villages were being deserted, with local communities struggling to keep their young folk as people left seeking employment and new opportunities on the coast. However, though economically poor and declining, the north is a wilderness of cultural and environmental treasures.

The Ministry and NTO launched the new initiative with a public forum as part of its master planning process. With hiking and biking recognized as popular sports, it was a natural choice to create new adventures for new markets, and so *Montenegro Wilderness Hiking & Biking* began. More than 500 stakeholders participated in a journey to find sustainable solutions that would link the coast with the hinterland and north, open up new geographic areas of the country, and create on-trend adventures for visitors to Montenegro that would benefit impoverished remote communities.

Stakeholders made a commitment to protect the integrity of these wilderness areas. A labyrinth of forgotten trails allowed for sensitive development of existing natural assets: paths used by Roman and Ottoman caravans as well as historical military trails that had linked Austro-Hungarian fortresses made ideal routes for hiking and biking.

More than 237,000 hectares of wilderness are protected in Montenegro, with 100,000 hectares of National Parks representing 20% of Montenegro's total territory. Many of the areas targeted for trail development were in truly remote areas of the country. Despite many challenges, the potential for the program remained ambitious in size and scale and today it connects six coastal tourism centers to 13 northern municipalities with a network of 6,000 kilometers of trails. To make the hiking and biking trails a reality, the government has invested more than one million Euros, and this investment has been supplemented by seven international partner organizations providing project-specific technical and financial support.

As a starting point, existing spot or regional initiatives were collected under one roof to create a nationally coordinated project. Since 2007, Montenegro has collected its own relevant market intelligence data. The latest study was conducted in 2011, to help fine-tune the approach according to potential guests" experience and expectations, which vary remarkably even in the European source markets.

Because the Trendscope pre-study delivered more than 80 travel motives, a selection needed to be made. For this selection, the Zurich Model of Social Motivation

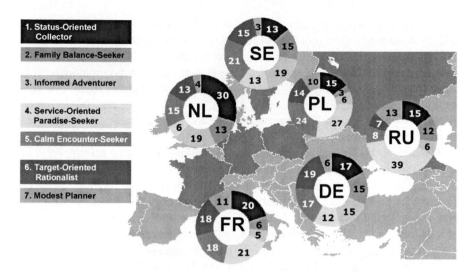

Fig. 7. Trendscope traveller types – distribution in Europe (2010)

conceived by Norbert Bischof provided the theoretical basis. In the second step, a comprehensive quantitative study was carried out in six European countries (Germany, France, The Netherlands, Sweden, Poland and Russia). Altogether, 6,000 respondents (1,000 per country) were interviewed via an online panel. In this way, the results of the preliminary study have been quantified and representative information on the target groups in important European source markets has been collected.

To finally define the Trendscope Traveller Types, an approved multivariate statistical analysis method was applied. The core element for establishing the typology was cluster analysis, a mathematical-statistical method for data reduction by groupings. On the basis of the pre-study, a substantial statement battery was composed, demonstrating all the relevant needs and motives linked with informational, booking, travel and after-travel behavior.

The statement battery was applied in the above-described quantitative study in six European countries. By means of a factor analysis the single motives were classified into motive dimensions and the statement battery was reduced to a straightforward, valid and selective set of items. Using cluster analysis, the participants were grouped into the Trendscope Traveller Types based on their similarities or dissimilarities. Each group is therefore as homogeneous as possible, whereas the different groups are as heterogeneous as possible.

The segmentation approach outlined above brings key advantages in developing and marketing Montenegro as a destination for activity holidays by choosing custom-fit marketing for special target groups in priority source markets. In the end, the investment in this work will pay off in the form of more readily targetable promotional activities and through the sharing of knowledge and recommendations; the returns will be faster and stronger.

The potential of hiking and biking and related activities for economic, cultural and environmental development is as spectacular as the diversity of Montenegro's offers for nature and adventure tourists. Already, there has been impressive business development, including specialized travel agencies, new accommodations to serve travelers, flourishing local gastronomy, domestic products, traditional souvenirs and contemporary handicrafts, plus all related hiking and biking enterprises: equipment rentals, repair services, bike and luggage transportation, as well as local guide services.

Montenegro Wilderness Hiking & Biking is a year-round program supported by diverse stakeholders and focused on directly benefiting local residents. Participating stakeholders include the Ministry of Tourism, the National Tourism Organization, the 21 participating municipalities with their Local Tourism Organizations, and a Regional Development Agency. The many participating special interest groups include the Mountaineering Association of Montenegro (PSCG), the Cycling Association of Montenegro (BSCG), the Mountain Rescue Service (GSS), and the management of the five Montenegrin National Parks. The many local communities with family businesses situated along the trails are at the heart of the success of *Montenegro Wilderness Hiking & Biking*.

Since its inception the program has achieved significant milestones:

- A comprehensive hiking and biking trail network of 6,000 kilometers was identified and registered electronically, and a good deal of infrastructure completed, i.e. marking, signposting, provision of weather shelters and mountain huts for accommodation.

- The government has provided the necessary supporting legal framework, with the new Mountain Trail Law and further regulations being implemented, plus development and implementation of new "Bed and Bike" (2010) and "Wilderness Camping" (2011) standards.

- Given the remoteness of many areas, security has been addressed and the capacities of the Mountain Rescue Service (GSS) reinforced with training and equipment; mountain guides have been further trained.

- The major issue of keeping the trails pristine has been addressed with a clean-up and education program: all garbage points along the network were identified and documented for action. This program has been set up with local action plans towards a sustainable system, as well as "Leave No Trace" and "voluntourism" aspects. On the national level, the supporting projects include ongoing awareness campaigns, clean-up activities and training, such as "Let It Be Clean" and "This Land Is My Home" – also providing new employment for 140 economically displaced, unemployed persons.

- There has been extensive capacity building and support for community development and support for cultural heritage with ongoing workshops, detailed handbooks, presentations and international exchange programs.

- 39 certified mountain guides have been trained. Training programs are both specific to servicing the hiking and biking market niche but also more generally geared toward helping people understand what sustainability means, and offering support in creating and protecting sustainable programs such as this one.

- This extensive transfer of industry skills and support for new product development has resulted in new local capacities, such as 15 specialized agencies. They link and connect new local service providers for creating promotable packages for hiking and biking markets.

- Print and online materials including maps, brochures, GPS data and a detailed "Wilderness Biking" road book covering 1,700 km of trails have been published.

- To build global recognition and a reputation for "Wilderness Hiking & Biking," as well as to support international sales, six familiarization trips for international tour operators and media have been organized, hand in hand with local specialized agencies and service providers for know-how transfer.

Montenegro Wilderness Hiking & Biking highlights tradition and heritage in traveler itineraries. The trails and associated activities make a feature of the old ways of living in the remote villages and "katuns," (traditional mountain pastures at high altitudes). These include preserving traditional building styles, rural life, making homemade produce ranging from dairy and meat products, to the authenticity of making local breads, honeys, herbs, wines and "rakija." Traditional handicraft is also encouraged, and artisans have been organized so they can sell to new markets. The following illustrates some ways in which the program has been successful for local inhabitants:

The implementation of the quality standard for cycling-friendly accommodations started in February 2010, and now 20 privately owned accommodations along the trails have been qualified with the new "Montenegro Bed & Bike" quality standards. One of the remotest family-owned "Etno selo" type of accommodations, called "Izlazak," opened in 2009 along the "So Scenic" Top Trail and has recorded 417 overnights in 2010, a 47% increase of business over the previous year. A very remote katun in the vast highland of Sinjajevina that was going to shut down had over 80 mountain bikers passing by its doorstep in 2010 and ensuring its ongoing viability perspectives.

The program has used the methodology of *Results Chain Analysis* for development, and has introduced disciplined annual monitoring. This helps to qualify and quantify the process, from identifying assets and opportunities, to developing and implementing activities, to capturing results. The *Global Sustainable Tourism Criteria* will soon provide a further set of relevant metrics. The following figures illustrate the first monitored cycle from January-October 2010 after three years of the program:

- New tour operators generated 3.2 million Euros in revenue for local businesses.

- 91,600 guests – out of a total of about 1.2 million – cited hiking or biking/cycling as their main activity in Montenegro (on individual and package tours).

- The number of guests, overnights, and revenues in the National Parks and the northern municipalities has increased significantly:

  — Number of guests in the northern municipalities: plus 20% (40,000 guests registered May-Sept)

  — Overnights in the northern municipalities: plus 27.5% (95,250 overnights in the same period)

  — National Park visitors: plus 20% (153,000), revenues: plus 16% (360,000 Euros)

- Average daily spending of hiking and biking guests is equal to the average spending on cycling trips in Western Europe: without arrival/departure travel cost, 64 Euros.

- Over 70 new specialized quality tour operators from European/US source markets have included Montenegro in their programs for 2011 and plans for 2012 from a zero baseline.

- For 68% of their clients, the average length of stay is 7–10 days.

- The tourist season has been extended: 68% sell April–June, 82% July–September, 25% October-December, 7% January–March.

Though using different words to describe it, Montenegrins have understood sustainability for hundreds of years – ever since 1878, when King Nikola proclaimed the country's first Nature Reserve. In 1991 the Parliament proclaimed Montenegro an Ecological State. More and more milestones have been achieved as Montenegro has continued to manifest being "eco by nature," whether managing projects in a sustainable way, reserving territories for nature protection, making policy, reviewing and adjusting strategic initiatives, aligning laws, regulations and compliance with international treaties and proclamations, or supporting ecological education and awareness.

Montenegro Wilderness Hiking & Biking is an initiative that opens up new experiences for visitors while preserving local traditions and improving the quality especially of rural life, encouraging young people to stay and to eventually return to revived village economies. In its third year, the program has contributed to achieving a number of strategic goals as detailed in the updated Tourism Master Plan (2008) and the Montenegro Tourism Development Strategy to 2020:

- Meeting new trends for higher yield, nature-based tourism

- A new integrated tourism offer, linking the coast with the hinterland and mountains in the north, using natural and cultural assets for creating new business

- Extending the tourist season and smoothing the flow of annual visitations

- Increasing revenue, and supporting new jobs and income opportunities, plus a host of soft benefits for local communities including nurturing and protecting authenticity, cultural treasures and traditions

- Exploring further adventure travel product diversification, based on the unique values of the natural environment

- A range of new and innovative local enterprises supported through comprehensive legal, institutional and infrastructural measures

It will take ongoing hard work to fully unfold the socio-economic and environmental potential of *Montenegro Wilderness Hiking & Biking*. The increasing popularity demonstrates its appeal for a whole new group of adventure travelers coming to explore Montenegro along the trails.

One of the latest breakthroughs is to use *Montenegro Wilderness Hiking & Biking* to facilitate cross-border activities with the neighboring countries, in formerly highly restricted border areas, for example with Albania. This promotes new jobs for local people in tourism and fosters cooperation. Apart from the official border crossing points, a simplified procedure has been established with the Border Police to ensure safe and easy passage for guests. Examples of cross-border cooperation starting in 2011 are the projects "Peaks of the Balkans," with Albania, Kosovo and Montenegro, and "Via Dinarica," connecting the Dinaric Alps region with a development corridor.

"Finding the right program to encourage integrated and sustainable rural tourism development and village regeneration was challenging. After analyzing best practices internationally, we were clear on the fact that any new program we wanted to invest in must have critical mass to generate sustainable job and business opportunities for local communities, while at the same time it must help protect the environment and preserve rural traditions and culture.

Judging by the enthusiasm and ongoing constructive action by all stakeholders, we are confident that *Montenegro Wilderness Hiking & Biking*, with all of its related elements, means nature-based adventure tourism is a key driver for balanced growth and prosperity for our country. Our approach and choices show what is possible when we all internalize the principles of becoming 'eco by nature' so we can truly achieve economic, social, cultural and environmental balance and sustainability."[55]

---

[55] Mr. Predrag Nenezic, Minister of Tourism 2001-2010.

5   Conclusions

"Strategies are the unique and sustainable ways by which organizations create value."[56] The route to results starts from a destination management organization's core purpose, and that route can turn out to be a rollercoaster ride. There is no standard approach to change, because no two destinations are identical. But some basic things are common to all, such as the need for focus, for thinking in results rather than activities, and the true meaning of excellent brand management and market intelligence.

Before starting a market survey, it has to be considered to what extent market intelligence shall be used. Far too often this is not the case, and market data is merely being administered, or just being used for convincing politicians that either the destination management organization is doing a good job and therefore needs further funding or that they should invest more money in the promotion of tourism in the future. Before conducting any surveys it should be made obligatory that the resulting information be used for the product creation and marketing process, and the knowledge shared with stakeholders, partners and media. Where such information is much too rarely being used is in the field of strategic tourism market development. When the main reason for market research – the gathering of knowledge – is not taken into account after having conducted a special market research campaign, the money spent for this campaign would be better invested elsewhere. However, not all destination management organizations are able to use market intelligence in the same way. Depending on the resources of a destination, it might be a good choice to completely abandon (primary) market intelligence and to save this money for other important tasks. To be able to do so, the destination must boast not only potential, but also a strong and sharing network.

No destination can function without the "we" – a destination management organization needs big and small partners, and those partners need a visionary, well-networked and personally trusted lead umbrella organization.[57] Cooperation and communication are bywords: in the great industry of travel we would do well to practice the spirit of *Ubuntu* and share what we know – what works and what doesn't work – for the good of all. But results never come overnight.

In a destination, whatever you do or don't do will become visible one day: you simply cannot "not communicate." Messages come in words and deeds – no sugar-coating necessary, because nothing supports successful change more than when the behavior of important individuals and institutions is consistent with their words.[58]

Experience proves that decision makers should go for the long tail by genuinely investing in the holistic thinking that links the destination's mission, values, vision and strategy with the target groups. Today's travelers have a huge selection avail-

---

[56] Kaplan/Norton: p. 3.
[57] Destination Camp 2011 results.
[58] Compare with Kotter, p. 13.

able, and unless they feel their values and convictions represented in each and every ingredient of a destination, they will choose and recommend another one; personal responsibility, credibility and building trust are the highest form of modern hospitality.

The case of Montenegro is a success story, and for readers not involved in every step, sidestep and all the daily struggles, it might be a compelling one to "copy and paste." However, there are plenty of valuable lessons that have been learned at all levels concerning management, customer orientation, facilitation of business, cooperations, financing/return on investment, sincerity and consistency. A profound analysis of those aspects would go beyond the scope of this article, but they are known and increasingly discussed in and outside the country, which again is a good sign.

A deep understanding of the ways this industry works and how it does *not* work is key to putting destination development in relation to the rest of the world: successful change is all about meaningful results, partnerships and creating an environment of purpose and credibility, in order to find "meaning beyond survival."[59] We will succeed as an industry if we do so. Destination "biotopes" that feel out their very own and specific blend of favorable conditions, and are able to nourish, combine and steer them properly, will grow and flourish.

References

Bischof, Norbert: *Das Rätsel Ödipus*, Munich 2001

Butler, R.W.: "The concept of a tourist area cycle of evolution: Implications for management of resources", in: *Canadian Geographer*, 24 1980: pp. 5–12

Collins, James C. and Porras, Jerry I.: "Building Your Company's Vision", in: *Harvard Business Review on Change*, Boston 1998, p. 30ff.

Dym, Barry; Egmont, Susan; Watkins, Laura: *Managing Leadership Transition for Nonprofits: Passing the Torch to Sustain Organizational Excellence*, New Jersey 2011, p. 20ff.

Hopkins, Rob: *The Transition Handbook*, Totnes 2008

Huttunen, Pekka: *Onnistuneen konsulttihankkeen toteuttaminen*, Helsinki 2003, p. 19

Kaplan, Robert S., Norton, David P.: *The Strategy Focused Organization*, Boston 2001

Kotter, John P.: "Leading Change: Why Transformation Efforts Fail", in: *Harvard Business Review on Change*, Boston 1998

Megginson, Leon C.: "Lessons from Europe for American Business", *Southwestern Social Science Quarterly* (1963), at: www.darwinproject.ac.uk

Müller, Mokka: *Das vierte Feld – Die Bio-Logik der neuen Führungselite*, Munich 2001, p. 30ff.

Porter, Michael E.: *Wettbewerbsvorteile: Spitzenleistungen erreichen und behaupten (Competitive Advantage)*, Frankfurt/New York 1992

Rifkin, Jeremy: *The Empathic Civilization*, Cambrigde 2009

---

[59] Rifkin, Jeremy, p. 39ff.

Rück, Hans and Mende, Marcus: "Innovations in Market Segmentation and Customer Data Analysis," in: Conrady, Roland and Buck, Martin (eds.): *Trends and Issues in Global Tourism*. Berlin/Heidelberg 2008: pp. 137–148

Schaffer, Robert H., Thomson, Harvey A.: "Successful Change Programs Begin with Results", in: *Harvard Business Review on Change*, Boston 1998, p. 189 ff.

von Förster, Heinz: Wissen und Gewissen: Versuch einer Brücke, published by Siegfried J. Schmidt, Frankfurt Main 1993, p. 49

World Economic Forum, *Travel & Tourism Competitiveness Index 2008*, "The Case of Montenegro: Steps toward environmentally sustainable tourism development", p. 32 ff.

World Economic Forum, *Travel & Tourism Competitiveness Index 2011*

Articles and Websites

www.adventuretravel.biz
www.bpic.co.uk/articles/deadhorse.htm
www.change-management.com
www.familien-ferien.de/Regionen-Orte/Schwarzwald/Ferienland-im-Schwarzwald
www.guardian.co.uk/travel/2007/jun/30/saturday.helsinki
www.rakvere.ee
www.suedtirol.info
www.tourismfortomorrow.com/Winners/Previous_Winners_and_Finalists/Destination_Ste
    wardship_Award_/country-of-montenegro/index.php
www.visitestonia.com/en/holiday-destinations/city-guides/rakvere
http://blog.kmto.de/veranstaltungen/die-marke-im-tourismus-ergebnisse-des-
    destinationcamp-2011
http://de.wiktionary.org
http://en.naturparkschwarzwald.de/regional
http://media.unwto.org/en/press-release/2011-06-20/tourism-linking-cultures-unwto-
    launches-world-tourism-day-2011

Interviews and Presentations

Interview with Mr. Johann Friedrich Engel, Managing Director Creatop GmbH, Germany, Advisor to Masterplan 2001/Update 2008, and to the Tourism Development Strategy to 2020

Interview with Mr. Predrag Nenezic, Minister of Tourism, 24th July 2001, in *World Investment News*

Interview with Ms. Anita Meier, Founder and Director of the Mallorca Information Office in Frankfurt 1982–89, consulting, destination management, marketing and PR, later also responsible for the Balearics

Ms. Nicky Fitzgerald, a luxury adventure travel industry and marketing veteran, speech at the Adventure Travel World Summit 2010 in Aviemore, Scotland

Mr. John Sterling, Executive Director, The Conservation Alliance, speech at the Adventure Travel World Summit 2010 in Aviemore, Scotland

Tourism Development International: Presentation 10th Feb 2011, ETC/UNWTO Handbook on Tourism Product Development

Others (Montenegro Case Study)

Montenegro Tourism Development Strategy to 2020
Pilot Compilation of Tourism Satellite Account in Montenegro for 2009, Draft Report July
    2011, p. 6
Press release of the National Tourism Organisation, March 2011
Statistics 2000–2010, made available by the Ministry of Sustainable Development and
    Tourism and the National Tourism Organization of Montenegro

Social Media and Technology Tackle Tourism Industry

Social Media and Mobile Devices

David Perroud

What role do social media like Facebook, Twitter or Tripadvisor play in the book-ing and travel behavior? Who uses them and for which purposes? Which changes do they imply for the travel industry? This paper helps answering these questions using a global survey, carried out by m1nd-set, a market research agency dedi-cated to travel research.

1   Research Methodology and Sample

Using a part of m1nd-set's database of 20'000 travelers, recruited at 30 of the world largest airports, m1nd-set interviewed 1'365 travelers about their usage of social media for leisure travel purposes. All respondents were randomly selected and invited to participate to an online survey.

Participants are mainly from Europe (48%), North America (30%) and Asia (15%). They tend to travel more frequently for business and to be more frequent travelers than the average traveler. The sample is balanced in term of all age groups (from 18 y.o.). All participants answered a set of questions about their leisure travel habits and preference that has been used to segment respondents via cluster analysis.

61% of the interviewees could be segmented in 3 groups of similar travel pat-terns (see figure 1). The first group "Adventurers" is made of travelers who can be best described by their willingness to sacrify on comfort for discovering new places. Adventurers are well traveled and feel at ease in foreign countries. The 2nd group, "Mainstreamers" is made of younger people, who like to go to well known places in familiar hotel chains, with many things to do, animations and plenty of other tourists. The 3rd, group, "Selectors" are somewhat older travelers, who take extra care choosing comfortable places. They favor boutique or exclusive hotels, like good food and, once they have found a place that meet their difficult criteria, love to come back.

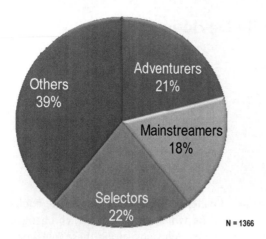

Fig. 1. Sample segmentation

Source: m1nd-set, research on social media

2   Usage of Social Media for Leisure Travel

Half of the travelers use social media on a regular basis to get information before booking a leisure trip. 20% use them for almost every trip. As expected, age plays an important role, since about 40% of travelers aged 55 and older do not use social media for information before booking. This proportion falls to 20% amongst less than 25 years old. Figure 2 below shows that Tripadvisor and Facebook are the two mostly used social media to get pre-travel information. Facebook is #1 for the "Mainstreamers". Other social media, including groups or fan pages from travel agencies, are much less used.

Whilst social media are frequently visited to get pre-travel information, they are not commonly used for booking trips. Only 6% of our sample claim to book through them regularly and 16% occasionally. Getting pre-travel information on social media is more common that getting information whilst travelling, still, around 37% say that they connect to social media to get relevant information during their trips. This number rises to 45% among Mainstreamers. That's a relatively high number, players of the travel industry targeting this segment of visitors should count with the fact that almost 1 out of 2 is seeking and exchanging information about their current trip on social media.

2.1   The Influence of Travel Recommendations/Reviews

Nowadays every destination, hotel, activity, restaurant is rated and commented on social medias. How does this influence booking behavior for leisure travels? 52% say they use travel recommendations, frequent travelers (68%), Mainstreamers

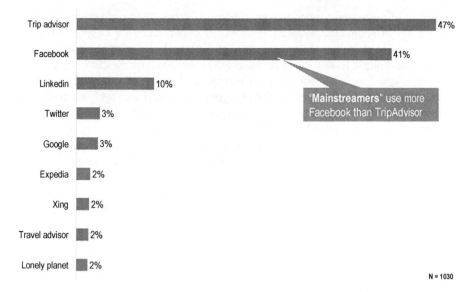

Fig. 2. Social media used

Source: m1nd-set, research on social media

(66%) and women (61%) are those who use recommendations most. A large proportion (68%) claim to be influenced by recommendations and reviews and this raises to around 75% for the same 3 groups mentioned above (Frequent travelers, Mainstreamers and Women). Additionally, around 2/3 of the sample say they give travel recommendations or reviews on social media.

2.2   Implications for the Travel Industry

It is no secret to reveal that social media brought a major revolution in the way travelers gather information about their destination, hotels, activities, etc. A majority of them will go on Tripadvisor or Facebook to get feedback from people who have had the experience before. As a consequence, the world has become smaller and what happens in a hotel at the other corner of the globe can be instantly known to internet users around the world.

The travel industry that used to have a unilateral control of information and content now has to learn how to manage information the same way a "neighborhood business" would, where word of mouth plays a key role. Sources of content are multiple and it is no longer enough to manage one or two channels, like one's own leaflet and website. Future customers expect much more like recent photos, video, posts from current and past customers, etc. Social media, like Facebook offer a unique, powerful and cheap opportunity to provide it, assuming one knows how to take advantage of them. The other and more difficult aspect to manage is a total shift of customers" voice power. If before a dissatisfied customers with, let's say a hotel in Katmandu, would tell about her experience to a couple of friends, now she might share her frustration, through web reviews, with hundreds of people. And to make it worse, the exact people who are considering this hotel for their next holidays. Since, as our research shows, a majority of potential customers read, write and get influenced by travel recommendations and reviews, managing customer satisfaction and complaints has become of outmost importance.

Those who understand how to best leverage social media, by using to their advantage the great information power they have, will get a significant competitive edge. It seems obvious and simple, but during our study we tested dozens of airline, hotel, destination, restaurant websites and only about 15% of them are using social media the way travelers expect (and among them not many big names of the industry). Moreover moving from a provider-centric business to a customer-centric one implies a lot of structural changes that many organizations have yet to undertake.

3   Usage of Mobile Devices

As shown in figure 3, mobile devices are mainly used when it gives a direct benefit to travelers, like faster checking in or boarding on flights. The usage of mobile devices is not yet frequent for trip bookings, this is partly due to the fact that many websites are not adapted to the small screens of smart phones and booking through them can be tedious. It is also worth noting than 18% of our sample simply did not have smart phones.

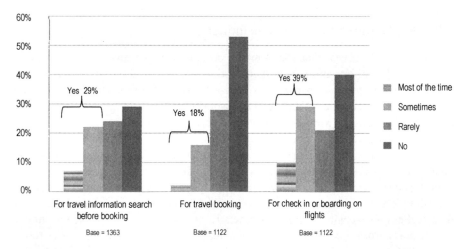

Fig. 3. Usage of mobile devices

Source: m1nd-set, research on social media

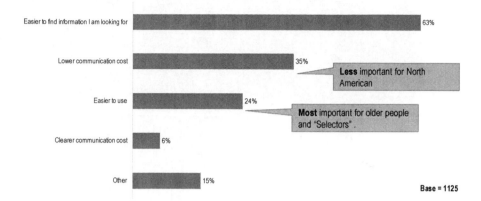

Fig. 4. What would increase usage of mobile devices

Source: m1nd-set, research on social media

19% use social media communities like Facebook for travel purposes on mobile devices and this is very age dependant since this number rises to 34% for the younger part of our sample (less than 35 y.o.). Roughly the same proportions use Facebook places while travelling. Figure 4 highlights what would make travelers use their mobile devices more for travel purposes.

Interestingly the biggest driver is not usage costs, like roaming but a more adequate content. Players of the travel industry who seriously want to increase mobile device usage have to provide tailor made content, not only to fit the format of smart phones but also to fit the spirit of an immediate, useful and located information.

4   Conclusions

The outcome of this first global research on social media and mobile devices usage for travel purposes shows real numbers for something most industry players were feeling: both social media and mobile devices play an increasingly important role on people's booking and traveling behaviors. The extent to which social media, through their content sharing power, are rocking over the tools travelers use to get content & information is somewhat surprising and, given the difference of between age group is bond to increase. m1nd-set will follow-up these trends regularly and is dedicated to help the travel industry adapt to these new media by monitoring trends and giving adequate workshops and trainings on how to best (re)act in this rocked-over environment.

Holiday Hotels and Online Booking Behavior – News from the GfK Research Panels

Alexandra Weigand

1 Methodology and General Background

1.1 Methodology

The following essay will start by showing how the market of tourism in Europe is developing at the moment. It will then analyse differences that exist between the booking behavior at travel agencies and in online sales. One of its features will be the scope of preferences while choosing a hotel during a stay. All of these analyses are based on the retail panel of GfK Retail and Technology GmbH – or shorter GfK Travel Insights – which has been available for several years now in Germany, the United Kingdom, Italy, the Netherlands, France, and Russia. Thanks to a representative number of travel agencies, who agreed to data supply, we are able to evaluate travel data week by week. Additional data from tour operators are implemented in many countries. In Germany, for instance, the sample includes over 1,200 travel agencies. Here, a collection of 80,000 entries is supplied every week, summing up to over four million bookings per year; – a respectable evaluation data base to draw on, and beyond this there is evidence for theories about booking behaviours in the tourist market.

1.2 General Background

While 2009 had been characterised by economic and financial crisis, coupled with a 4.7% decline in GDP, clear signs of a recovery in Europe appeared by the end of the year 2010. Germany managed to impressively overcome the crisis: with a 3.6 percent increase in GDP for the first time after the Reunification. There are also similar promising developments in Sweden and in Poland, and France, the United Kingdom and Italy managed to recover well. All in all however, a North-South divide appeared during the economic recovery; for example Spain couldn't match last years GDP. This decline is also reflected in the development of spending in the tourism business.

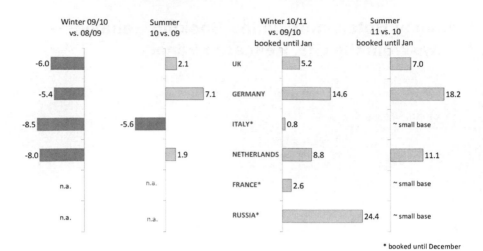

Fig. 1. Sales value growth +/– in %, booking development

Source: GfK Travel Insights

2 Booking Trends in Europe

Figure 1 shows the current booking trends in turnover in the six abovementioned source countries. The 09/10 winter season was heavily influenced by economic crisis. Turnovers collapsed in all countries, but Germany managed to escape with the smallest turnover loss, just 5.4%. In the summer season 2010 that followed, clear signs of recovery emerged, thus the Netherlands and the United Kingdom were able to top their annual results of the previous year by approximately 2% and Germany even by 7%. Only Italy has been suffering from the repercussions of the economic crisis for some time now, lagging behind the summer of the previous year.

Per current level of bookings at the end of January 2011, the winter season 2010/11 and the summer season 2011 show good results: All countries surpassed their turnover levels of summer 2010. Yet, as mentioned before, the opening of a north-south divide can be observed simultaneously with the GDP trend. Countries more to the south, such as Italy and France, can book only a slim turnover gain of 0.8 or 2.6% respectively. At the beginning of the year, summer 2011 comes up with a record alert for Germany as turnovers go up by 18.2%. It remains to be seen whether this advantage can be held through the season. Forecasts for the summer season for Italy, France and Russia cannot be made in January, as travel bookings in these countries are arranged definitely at a later date.

In all source markets which are monitored by GfK, Spain is most frequently preferred as a holiday country: Spain's isles and mainland achieve the highest turnover shares at travel agencies. Second and third place is occupied by Turkey

Table 1. Top 10 destinations by country; summer 2011 vs. summer 2010, sales value %, growth +/– in % booked until Jan 11

UK	Value %	Growth %		GERMANY	Value %	Growth %		NETHER-LANDS	Value %	Growth %
TOTAL	100	7		TOTAL	100	18		TOTAL	100	11
1 SPAIN	24.5	16.2	1	SPAIN	24.4	22.6	1	SPAIN	17.7	13.6
2 TURKEY	12.1	3.4	2	TURKEY	18.8	22.5	2	TURKEY	15.5	27.8
3 GREECE	11.3	10.6	3	GREECE	7.1	-12.0	3	GREECE	13.6	8.5
4 USA	8.9	4.2	4	USA	4.4	12.7	4	ITALY	6.3	7.4
5 CYPRUS	5.4	17.3	5	EGYPT	4.2	-10.2	5	FRANCE	5.4	0.0
6 CARIBBEAN	5.0	-12.2	6	ITALY	3.8	28.1	6	USA	5.0	18.7
7 EGYPT	4.1	-7.3	7	GERMANY	3.5	45.0	7	PORTUGAL	4.1	9.9
8 MEXICO	3.7	19.8	8	BULGARIA	2.4	20.1	8	EGYPT	2.9	3.7
9 ITALY	3.1	15.6	9	TUNISIA	1.8	-15.4	9	NL ANTILLES	2.8	12.9
10 PORTUGAL	1.9	6	10	CANADA	1.4	33	10	INDONESIA	2.2	-3

Source: GfK Travel Insights

and Greece respectively. Greece is gaining considerable turnover increases in the Netherlands and also the UK, while German holidaymakers believe that Greece has not yet recovered from its political crisis of last year. A further high-ranking destination is Cyprus, which shows a positive trend for this summer according to holiday-makers in the United Kingdom. The destinations U.S.A., Italy and local resorts are becoming increasingly attractive for German travelers. Our Dutch neighbours don't just book theirs holidays in the U.S.A. and Italy, which are fast growing destinations, more frequently, but also increasingly travel to the Netherlands Antilles.

Destinations in northern Africa are very much affected by political turmoil. Bookings for Egypt and Tunisia ebbed away in the second calendar week to the benefit of other destinations, such as Turkey, the Canary Islands and the Balearic Islands. Since the seventh calendar week however, bookings for Egypt seem to be gradually recovering.

3   Booking Behavior Online vs. Offline with Focus on Hotels

Since the beginning of 2011 the surveys conducted by GfK Travel Insights in Germany also include, apart from bookings in the stationary distribution, the bookings arranged online via internet platforms. The Online Panel is still too young to predict the actual size and development of online distribution. Nevertheless a comparison between both booking channels reveals interesting differences even today in the booking behavior. A few hypotheses will provide a means to expound on these differences. The analyses are based on the bookings of approximately 20 travel portals in January and February 2011.

Hypothesis 1: Families prefer booking at travel agencies

Figure 2 establishes that the turnover share gained by online booking for a holiday with more than three persons – inclusive families – covers 38%, which is slightly above the share of travel agencies (37%). Therefore, hypothesis 1 cannot be confirmed.

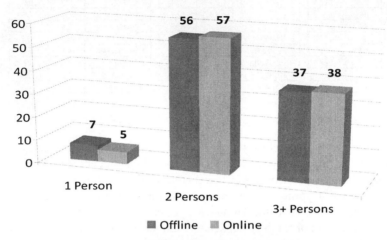

Fig. 2. Sales Euro % of number of persons, bookings until week 8/2011

Source: GfK Travel Insights

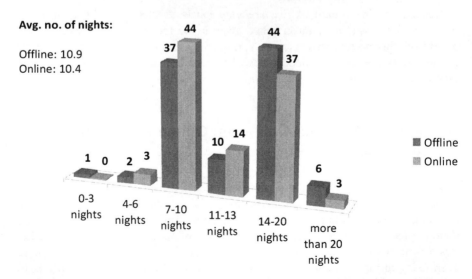

Fig. 3. For the main holiday travel agencies are preferred, sales Euro % of duration classes, bookings until week 8/2011

Source: GfK Travel Insights

Hypothesis 2: Bookings for the main holiday are preferably arranged at travel shops

Figure 3 gives an overview of duration classes and their shares in the online versus stationary sales.

Longer travel is definitely more often booked at travel agencies than on the Internet, while short trips of four to six nights are more likely to be booked online. At least it can be established that longer journeys are mainly booked at a travel agency.

Hypothesis 3: Last minute travel tends to be booked online

A further difference between stationary and online booking lies presumably in the time span from booking to departure. Is it true that last minute travel is booked preferably via the internet? Figure 5 provides an answer to this.

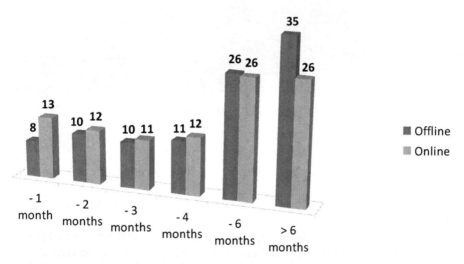

Fig. 4. Last minute trips are booked online, sales Euro % of prebooking classes, bookings until week 8/2011

Source: GfK Travel Insights

As the time between booking date and travel date gets longer, the more important the travel agency will become. Last minute trips account for 13% in total turnover gained by the online business, while they cover only 8% at travel agencies.

Hypothesis 4: Online customers are more concerned about prices

A pivotal factor for travel costs is the hotel category chosen. Hotel categories are not easily compared from destination to destination. In Spain, for example, two thirds of the total turnover is generated by hotels in the 3.5 to 4 stars class, but in

Fig. 5. Shares of hotel categories in four selected holiday regions.

Source: GfK Travel Insights

Turkey the 4.5 to 5 stars class is responsible for over one-half of all turnovers generated.

Figure 5 shows a comparison of the shares that hotel categories actually have in four selected holiday regions.

Across all destinations, however, it can be established that the share of simple hotels in the up-to-3-stars segment plays a more important role on the Internet.

With the Maldives as an example it can be noted that the luxury segment from 5.5 stars and over only plays a part in stationary sales. The elevated hotel segment between 4.5 and 5 stars has higher shares at travel agencies. The only exceptions are trips to Turkey. There, lower hotel classes are generally more important, but the 4.5 to 5 stars segment is still more strongly represented in the Internet than in stationary bookings. At any rate, it can be concluded that products of higher quality also meet a demand in the Internet sales.

Hypothesis 5: Long distance travel is preferably booked in the stationary business

Figure 6 gives an overview of the most important destinations in stationary and online distribution and their respective turnover shares. Yet, when we look at the

Country	Offline	Online
Spain	35,5	36,3
Turkey	18,3	25,2
Greece	7,8	8,4
United States	3,1	0,3
Bulgaria	2,5	2,5
Dom. Republic	1,9	2,4

Fig. 6. Long distance trips are booked offline, sales Euro % of destinations, bookings until week 8/2011 (including flight)

Source: GfK Travel Insights

Dominican Republic and the U.S.A. as long-haul destinations we receive a different picture. The Dominican Republic is a classic destination for beach holiday package tours and it has more sales occur over the Internet than in travel agencies. Trips into the U.S.A., in contrast, which are often composed of travel modules and include car rental, camping or round trips, are more frequently demanded at travel shops. With regard to the hypothesis it can be established, that it is not long distance in general, but complex arrangements which imply that consulting services offered by travel agents will be more readily appreciated.

Business Travel and Event Management

Ancillary Revenues in Air Transport – Gain and Pain

Stefan Fallert

1 My Very Personal Experience Lately

In March this year, for a long time again, I had the "pleasure" to fly domestic within the US. The flight from Los Angeles to Miami was scheduled for 08.00am. My American colleagues made me panic a little bit as they told me to be at least 2 hours at the airport before departure. So the night was short, I was extremely tired and not really "operational" when I arrived at the airport short before 06.00am. Well aware that I was be going to be charged US$ 25.- for my checked baggage which I could not pay upfront while doing the online check-in, I had my credit card already handy when approaching the check-in counter to drop off my suitcase. The lady from the airline was very nice and smiled at me, telling that I had seven pounds excess baggage. She kindly proposed to rearrange my suitcase and get the seven pounds out or to pay $ 65.- excess baggage fee. As said, being tired, having had no coffee and not really being able to have clear thoughts how to rearrange my personal belongings, I paid $ 90 in total for the two fees. To be fair, the flight was only around $ 200 incl. taxes but still I had to add almost 50% to the initial fare to finally get to Miami.

This personal story shows what passengers face day to day at airports around the world.

2 Ancillary Fees – A Great Alternative to Generate Revenues

2010, ancillary revenues summed up to more than 18 billion Euros according to the IATA – with a strong increasing tendency in the coming years (Source: Amadeus Guide to Ancillary Revenue by IdeaWorks, Oct. 2010).

The same source named "ancillary revenue champs" (including Allegiant, Flybe, Ryanair, Spirit, and Tiger Airways) generated an estimated 20% of total revenues through ancillary fees, while major US airlines were estimated at approx. 7% and "traditional airlines" at approx. 3%.

Although the travel industry accounts the introduction of ancillary airline fees to Ryanair, as this airline started to unbundle the air pricing, mostly US carriers have adopted this strategy and show how easily and quick additional revenue can be generated. But still Ryanair's CEO Michael O'Leary reconfirmed in an interview for Bloomberg Businessweek in autumn 2010 that he still favors the idea to give airline tickets for free and generate all company's income exclusively from ancillary fees (ancillary fees, on-board sales, commissions from selling insurances, hotels and rental cars through the Ryanair homepage).

This idea might sound wild for many classical flag carriers but also mirrors the mutual consent of many suppliers to maintain and increase revenues from those ancillary fees.

Although mainly US travel managers face the overall issue of unbundled pricing, this topic will be on the plate of their European colleagues sooner or later.

But with all fairness, let's first have a look at the overall market and developments in the past years:

Over the past decade, the airline industry had only three years of overall profitability. 2010 recorded a $ billion 16 net profit, compared to a $ billion 4.6 net loss in 2009. The profit margin (EBIT) might reach a 5% which is still below the average margin at the end of the 90s. The 2011 forecast now estimates a US$ 6.9 billion net profit, revised from a US$ 4 billion profit forecast in June 2011. 2012's first estimation has been set to US$ 4.9 billion, published in the September 2011 financial forecast report.

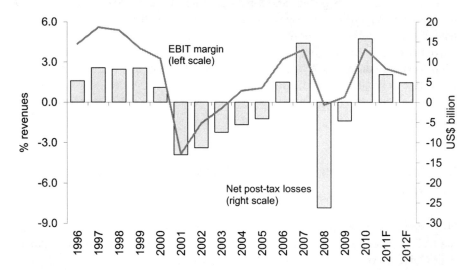

Fig. 1. Global commercial airline profitability

Source: IATA Financial Forecast, September 2011

There are several good reasons why airlines started unbundling their pricing:

- Internet and fare search engines have strongly increased the transparency of airline pricing. Led by Low Cost Carriers, the inflationary spiral put pressure on the pricing for all airlines. Face value of the airline ticket has become the main driver for customer purchasing behavior.

- Oil and jet kerosene prices remain high and directly impact the airlines profitability on a daily basis.

- Airlines promote the idea that unbundling reflects an advantage for the client as they will only be charged for the service they choose and use. No check-in luggage or no advanced seat options are "credited" with no fees.

- Ancillary fees represent a high revenue potential with extraordinary margins. As the sole selling platform for the time being are the airline's homepages or direct selling at airport, fees can be adjusted any time with no cost to the airlines.

- Airlines also promote the idea that ancillary fees increase the transparency. This might still need to be proven.

In general, the industry has introduced 3 main types of fees:

"A la carte"	Commission-Based*	Frequent Flyer Programs

■ Rebooking fees
■ Onboard sales (food, duty free)
■ Luggage (checked-in/additional)
■ Seating
  (pre-seating, emergency exit, etc.)
■ Boarding (priority boarding)
■ Booking/services by phone
■ Credit card fee
■ Lounge access
■ Internet
■ Onboard entertainment
	■ Hotel sale
	■ Rental car sale
	■ Insurances
	
	* through partnerships
		■ Selling/redemption of miles
		  (hotels, rental cars, other products)
		■ Credit card (co-branding)

Fig. 2. Major types of ancillary airline fees

3   Travel Manager's Main Challenge: Tracking

The concept of unbundling airline pricing has its pros and cons. Some travel managers might appreciate this approach, others won't. What all of them will not appreciate is the lack of data sources today to track and monitor these ancillary fees as part of their company's travel expenditure.

There is no yet approved industry standard; airlines have adopted the ancillary fee "strategy" in different ways.

The reasons for the tracking challenges are multi-fold:

- Lack of harmonization:

  Airlines have not agreed on a common pricing scheme. The industry estimates around 100 different fees on the market, pricing can potentially
  change every day, by airline, by markets. If you closely follow industry
  news, there is not one single week where you cannot read about the introduction of new fees and/or the increased level of implemented fees.

  To make it even more complicated and intransparent, many airlines have
  started to re-bundle fees into specific packages that might include advanced
  seat reservation, lounge access and priority boarding or checked-in luggage
  together with a meal onboard.

- Point of sale:

  Extra services can be booked and purchased any time throughout the lifetime of a trip: during the reservation or ticketing, during online check-in, at
  the airport or during the flight.

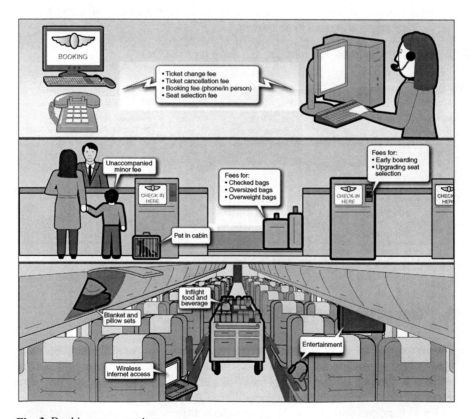

Fig. 3. Booking extra services

Source: GAO, US Government Accountability Office

Concentrated Might in the Sky – Airline Alliances and Travel Management

Oliver Graue

1 Introduction

Star Alliance long considered Central America as one the last unchartered territories on the world map. This is no longer so: With Copa from Panama as well as the Columbian-Salvadorian airlines Avianca and Taca three airlines from this region are joining the large airline alliance. In return the alliance is anxious about losing the Brazilian airline TAM, which has been a member of Star Alliance since May 2010. Its fusion with the Chilean LAN – once its fiercest rival of all things – could again put its membership into doubt, as the latter is part of the OneWorld alliance.

However it doesn't matter if it's TAM, LAN, Copa, Avianca, Taca or others: for the past few years, concentration in international air traffic has undergone an immense increase. By now three large airline alliances are ruling international air traffic. Star Alliance, Skyteam and OneWorld account for approximately 80% of all flights worldwide. A new power is rising in the sky: something like an oligopoly, which doesn't just turn its attention to "normal" leisure passengers, but also and especially to travelers from the business world. To cut costs in travel management, which in many companies has a great influence on the entire value-adding chain; travel managers generally negotiate special rates with airlines for their business travelers – so called Corporate Net Rates. However negotiating will be become more and more difficult, the more negotiating partners on the service provider side join together and form groups and the more the air traffic structure shows oligopoly traits.

Basically several developments lead to a continuous consolidation of the airline market:

1. Fusions and takeovers (for example: Lufthansa buys Austrian, Swiss, Brussels Airlines; in the USA: Delta Air Lines takes over Northwest; in Europe fusion of Air France and KLM and subsequent acquisition of parts of Alitalia).

2. Small carriers experiencing insolvencies and economic problems (for example: insolvency of the Lithuanian National Carriers Fly LAL).

3. Despite of a beginning economic upswing the trend for many airlines has been to cut out unprofitable routes and reduce further capacities.

4. Increasing number of codeshare agreements and other cooperations between airlines.

5. Boom of the three large airline alliances. This article focuses on this topic.

2   Current Airline Alliances Boom

Star Alliance emerged as the first global airline alliance and was founded in 1997 – followed by Oneworld in 1999 and Skyteam in 2000. However the alliance's great appeal to date to independent airlines has only unfolded in the past two to three years. Along with the Brazilian TAM, the US-American Continental Airlines, the Belgian Brussels Airlines and Aegean from Greece have joined the Star Alliance headed by the German Lufthansa. And in the upcoming months Ethiopian Airlines, Avianca, Taca and Copa will have followed by mid/end of 2012. Jaan Albrecht, former CEO of Star Alliance, does after all speak about a "great interest from unattached airlines". Air India, another candidate, will not be taken on board. The state-run operator had not met the minimum joining conditions.

Both competitive alliances Oneworld (formed around British Airways) and Skyteam are experiencing a similar development. Even though with the Russian S7 Airlines, formally known as Siberian Airlines, only one new member has been lately added to Oneworld, a list which names airline partners expected to join by 2012 includes heavyweights such as the German Air Berlin along with its Austrian affiliate NIKI, the Indian Kingfisher as well as Malaysia Airlines.

In turn Skyteam, which at the heart is formed by the company Air France/KLM, are expanding especially quickly in Asia. Vietnam Airlines joined along with the Romanian Tarom, Kenya Airways and Air Europa from Spain and five of the overall six candidate airlines will have originated from Asia by 2012. Next to Aerolineas Argentinas new admissions are planned such as China Eastern, the Taiwan China Airlines, Garuda Indonesia, Saudi Arabian Airlines and Middle East Airlines from Lebanon.

According to air traffic experts the reason for the increasing popularity of alliances is due to their attractiveness. "Above all airlines benefit from these alliances", airline advisor Andreas W. Schulz summed up at the ITB Business Travel Days 2011: "Being in an alliance enables them to offer passengers destinations as well as feeder and connecting flights respectively which aren't included in their own flight-schedules."

Yet also for reasons of cost airlines are increasingly joining alliances. Reasons for joining alliances includes joint sells offices and shared technologies as well as purchase discounts from suppliers, a key account management, shared use of airport lounges and low expenditure for frequent flyers and corporate incentives. Such opportunities are seen as clear benefits especially in times when margins of many companies are shrinking rather than growing.

Table 1. Airline alliances and their market share

	Star Alliance	Skyteam	Oneworld
Market Share	34 %	27 %	18 %
Members	Adria Airways, Aegean Airlines, Air Canada, Air China, Air New Zealand, All Nippon Airways, Asiana Airlines, Austrian Airlines, Blue1, BMI, Brussels Airlines, Continental Airlines, Croatia Airlines, Egypt Air, LOT, Lufthansa, SAS, Singapore Airlines, South African Airways, Spanair, Swiss, TAM, TAP Portugal, Thai Airways, Turkish Airlines, United Airlines, US Airways	Aeroflot, Aeromexico, Air Europa, Air France/KLM, Alitalia, China Southern Airlines, Czech Airlines, Delta Air Lines, Kenya Airways, Korean Air, Tarom, Vietnam Airlines.	American Airlines, British Airways, Cathay Pacific, Finnair, Iberia, JAL, LAN, Malev, Mexicana, Qantas Airways, Royal Jordanian, S7 Airlines.
Candidates	Ethopian Airlines, Avianca, Taca, Copa	Aerolineas Argentinas, China Eastern, China Airlines, Garuda Indonesia, Saudi Arabian Airlines, Middle East Airlines.	Air Berlin, Niki, Kingfisher, Malaysia Airlines.
Countries	181	169	145
Destinations	1,167	898	901
Passengers	604 million	384 million	335 million
Daily departures	21,000	13,100	8,750
Number of aircraft	4,025	2,225	2,012

Source: Data by Airline Alliances, Market Share by IATA.

3   Consequences for Travel Management and Business Travelers

A clear distinction of consequences which result from forming alliances must be made by business travelers on the one hand and by travel managers on the other. Both groups are driven by quite a different set of motives: the traveler desires a journey which is as quick and comfortable as possible, whereas for the travel manager cost issues stand at the forefront. VDR (Association of German Travel Management) business travel analyses always come to the result: Cutting travel expenses is by far the top priority for travel managers. Consequences are just as different as the points of view: Business travelers feel it is pleasant to have access to lounges in all airports, not having to change terminal when catching flight connections (airline alliances try to pool all members at one terminal) or even collect "valuable points" from all types of companies as part of frequent flyer programs so that desired premiums can be quicker acquired.

It is a different case for travel managers. Benefits such as being able to book flights via homeland airlines to destinations which are not approached by these themselves, but by alliance partners are offered to companies, for whom travel managers take on a representative role when dealing with travel matters.

However critical voices are increasing due to the high concentration, which forming alliances inevitably bring about. At the ITB-Business-Travel Days 2011 for example travel agency owner Otto Schweisgut (Isaria Lufthansa City Center Munich, LCC24) talks about increasing complaints when customers book a particular company due to quality reasons and then have to fly with an entirely different airline due to codeshare agreements. The travel expert sums up: "As much as airline alliances bring solidarity and standardization to the table, so little do they adapt in reality to service and quality".

Another issue: Company clients can't chose particular airlines anymore, with which they negotiate corporate net rates, but have to sign a joint contract with all members of the alliance – according to the motto "make or break". The supposed benefit – namely creating a wider variety – is when taking a closer look the biggest disadvantage: Rankings naming preferred airlines to date, with which travel managers like to work with when organizing business trips, would all of sudden be shaken up. In this context consulting firm Advito, BCD Travel's subsidiary, expects an increasing watering down of travel management in its business trip prognosis. Thereby suitable joint venture contracts could lead to more airlines being included in the travel management program than necessary. The consequence of this would be that business travel managers cannot fulfill negotiated capacity agreements with different companies because travelers have access to the entire portfolio of an alliance, which means that they can book a flight with an airline from a competitor with which a net corporate rate has been agreed. Jörg Gerhardt travel manager of the Munich chip producer Infineon at the ITB Business Travel Days 2011 hit the nail on the head when asking: "Why should I enter into contracts with 30 different airlines when I only need one?" Other travel managers even unofficially said that they would only be allowed to enter into desired contracts with an airline alliance if they would at the same time contractually tie themselves (unwanted) to the home carrier.

Concrete problems could emerge for travel managers during purchase negotiations when determining corporate net rates. Of course it is far more difficult to attain savings in near-oligopoly or even monopoly structures than in a market where a lot of competitors can be found. For the airline industry this would mean: Naturally it is easier to negotiate better rates on routes which are served by two or three competitive airlines rather than by connections, even though being operated by two or three airlines, that belong to the one and same alliance which perhaps even uses a single key account management. From experience this also applies to routes which are in the hands of two or even all three airline alliances. The best example is the transatlantic air traffic between Europe and North America. Although three alliances compete on this route U.S. Census Bureau figures show that ticket prices here have grown at a greater rate than routes which are additionally operated by "independent" carriers. Such oligopolies, though having a competitive relationship, seem to have a negative influence on price development, which however can be observed long ago in other examples from the global economy – similar development of fuel prices at filling stations springs to mind.

Of course airline alliances have a different view. Their argument is: Worldwide valid alliance contracts are of benefit to corporate clients as these buy less capacity for single routes and instead focus more strongly on entire regions. Alliances are more able to cover these than single airlines.

By the way airlines themselves are dependent on mighty oligopolies: This applies to the aircraft manufacturer market which Airbus and Boeing basically divide between themselves, as well as the oil industry.

4 Conclusion

Be it airline alliances, joint ventures or code sharers – experts recommend the same thing: Companies should look for individual contracts with airlines and demand these, advice which is for example also given by Advito. Basically observers assess the current phase of consolidation of the airline market as being quite insecure. For this reason alone travel managers should keep their options open.

Though competition will remain despite of consolidation through takeovers, fusions and alliances – many doubt whether it will continue to be as fierce as it is now. A positive for travel managers is that on the one hand the three alliances are in competition and on the other hand competition within alliances is (still?) so fierce that alliance members still entice clients away from one another: Each carrier pursues its own business goals. Despite of announcements made by some alliances to demand joint contracts in the future, terms and conditions of members are so different that agreements with two or three companies are practically impossible. The current trend to raise fees for services such as luggage transportation, meals, inflight entertainment, credit card payment or kerosene surcharges may probably make this even harder: There is a lot of chaos even within alliances.

Thirdly there are a number of independent carriers, which are primarily airlines from the Persian Gulf, which of course have greater financial funds than most other market participants. And finally some observers believe that some sort of low-cost segment will emerge for long-haul flights, which has now long become the norm for short and medium haul flights. This would lead to a stern competition, even for the mightiest alliances – yet whether an offer of such no-frill flights are suitable for successful business remains to be seen.

It is certain: Negotiating rates especially for transatlantic flights will become tricky in the future. This isn't to be expected for Asia due to numerous market participants, even though prices here don't increase any less, which however can be traced back to other reasons (not enough capacities, boom-region China). Experts are expecting a continuous consolidation in Europe – be it through takeovers or by joining alliances.

German companies beginning to copy the airline alliance model by forming purchase alliances is another consequence resulting from consolidation on the side of service providers. Here different companies combine their flight capacities and thereby confront sellers by pooling their purchase power. The number of such alliances on the side of travel managers is still rather limited. At the ITB Jörg Gerhardt travel manager of Infineon announced that they had just formed such an alliance with optics company Carl Zeiss and five smaller companies. The idea is a good one; the challenge will surely be not to just negotiate a joint contract but to collectively stick to it for at least a year. This however only works if travel managers have a strong mandate in their company.

References

Allianzen, Fusionen, Übernahmen. From BCD Travel in Motion, online, without place or date

Bartu, Friedemann: Wem nützen Airline-Allianzen? From: NZZ Online from 13th November 2009.

Graue, Oliver: Die Airline-Allianzen boomen. From BizTravel 3/10 from 16th June 2010, Hamburg, page 38 ff.

Graue, Oliver: Geballte Macht. From fvw 8/11 from 18th April 2011, Hamburg, page 46ff.

Graue, Oliver and Jürs, Martin: Von zu viel Macht und zu viel Unsinn. From fvw 6/11 from 16th March 2011, Hamburg, page 62ff.

Panel discussion "Contracts with alliances – pros and cons" at the ITB Business Travel Days 2011. Participants: Jörg Gerhardt (Travel Manager Infineon), Otto Schweisgut (Isaria Reisebüros Munich), Andreas W. Schulz (airline advisor CAT). Moderation: Oliver Graue (BizTravel), Berlin, 10th March 2011

Qnigge in Event Management

Markus Weidner

1 Introduction

Quality is a term which is used a lot, especially in industrial production and more than ever in the service sector.

During presentations at the ITB MICE Day on 9th March 2011 the speakers Markus Weidner, CEO of the consulting firm Qnigge® GmbH – Freude an Qualität and Susanne Frohreich, project manager of the agency Codiplan Gesellschaft für Concept, Dialog und Planung GmbH took a closer look at what quality means in the MICE industry. Along with taking theoretical approaches, the speakers tried to get the audience involved by starting an open dialog about quality in event management.

The question how event planners and service providers together could manage to thrill their customers through service and quality and make sure that a positive return of investment can be achieved was discussed in the course of the theoretical part.

During the practical part, which gradually developed throughout the entire presentation by constantly interacting with the theoretical part, a discussion between the speakers and the audience arose. In the course of this the interactive communication system "Q[kju:]-Interactive" of Codiplan was used. The system which was developed for events by Codiplan themselves is used for all types of events. It enables quantitative as well as qualitative data and opinions to be collected. For example compared to a TED (Total Electronic Designed Solution) system it can ask about the motives behind decisions which have been made.

For this presentation all of the approximately 160 participants were provided with a laptop and thus were able to give their opinions on questions about quality asked by the facilitator. During the presentation participants also had the chance to directly ask the speakers written questions about the topic using system. Results were presented on a screen for all to see.

2 Participants in the Auditorium

First of all it was established which trade affiliation the audience was composed of using "Q-[kju]-Interactive". Ninety-six of the 160 participants gave an answer (60%). The question was: "Please state which business sector you are from".

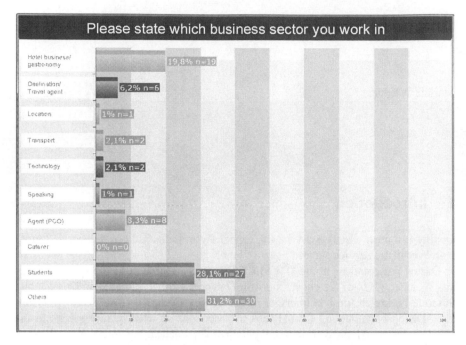

Fig. 1. Trade of business tabulation

Source: Codiplan GmbH

3   Defining Quality

Literature provides extensive and page-long definitions of the term quality or service quality (cf. Bruhn, 2011, p. 18ff)

This article will forgo a detailed and scientific discussion about quality. However the following must be said to understand the term:

At heart perception of quality of a service is rather subjective and can differ depending on the observer's point of view. Whoever takes part in a conference for the first time and as part of the catering service sees an adequately decorated table may, compared to an event expert who has seen a lot of fancy decorations, perceive this in a positive way. Whereas the experienced event participant may think: "I have seen this 100 times, that's nothing out of the ordinary".

Important for evaluating quality is therefore on the one hand the customer's demands (specifically stated by the customer) and furthermore his expectations (these are generally not stated in detail) and finally how these are met by the service provider or even surpassed.

If a customer expects a hotel room, which he has booked during an event, to have a neat and tidy bed, then this will naturally become what he wants to find and experience.

If he explicitly books a double room with a balcony, then he will of course, should he have managed to get this, perceive this in a positive way, but not as an extraordinary service experience. The room will only be perceived as being something special if the room provides service features which he did not order (request) or expect.

Features such as being able to connect to the internet subject to charge were once seen as an unexpected service, yet it is now taken for granted to find rooms with free internet access via wireless LAN or even the TV.

4   The Auditorium's Expectations

Using "Q[kju:]-Interactive" we were able to establish through a second survey what the audience expected of the presentation. One hundred and fifty expectations were given by the 160 participants. Multiple answers were possible.

The question was: "What do you expect of this presentation"?

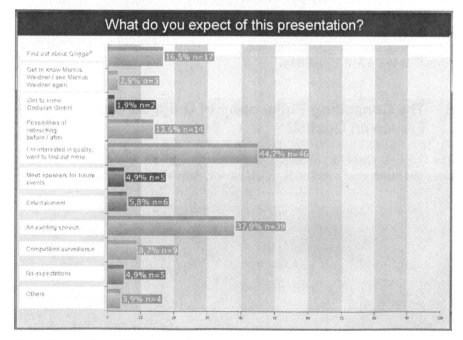

Fig. 2. Expectations of the presentation

Source: Codiplan GmbH

Next to interests concerning what quality is all about (44.7%), an exciting and interactive presentation topped the list (37.9%). Not having any special expectations was stated by 4.9%.

5   Explaining the Term Qnigge®

16.5% of participants, who took part in the second survey expected to find something out about the company Qnigge®.

Qnigge® is a fabricated word and stands for

- The company name of Qnigge® GmbH – Freude an Qualität (enjoying quality)
- As well as the concept of consulting and
- *The Qnigge® GmbH – Freude an Qualität* company values

Of course Q stands for the term quality and the phonetic word [K]nigge stands for Baron Knigge, who wrote the bestseller "On human relations" in 1788. In this book he already described the philosophical fundaments for service quality and explained in a unique way how people should behave towards each other and communicate to get along. Just his travel reports are worth reading and already describe how travelers back then perceived quality. (cf. Knigge 1977, p.268ff)

The Qnigge® idea was developed based on these by now old ideas combined with the modern service and process world. In hardly any other sector of industry are so many occupational groups needed to produce the end product that is an "event" as in the MICE-Industry.

6   The Consulting Philosophy of Qnigge® GmbH – Freude an Qualität

The Qnigge® GmbH's philosophy is characterized by six terms: Analyzing processes and management and employee behavior, defining and documenting company

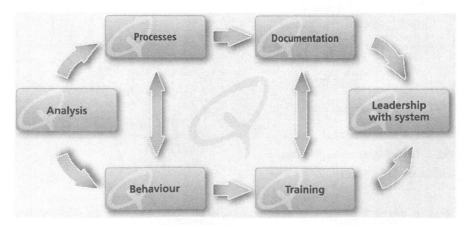

Fig. 3. "Qnigge® GmbH- Freude an Qualität" – consulting philosophy

processes, which are ultimately available as training modules and with the contents of the Qnigge® organization pyramid (see Fig. 4), to strive for a structured management.

By using this as a basis every service company can analyze its quality management and further develop it.

7 Introduction into Quality Management

Every successful company in the market has one way or another some sort of quality management. This is generally more or less formal, systematic and transparent. Here different management and controlling tools can be used (cf. Fig. 4).

During the past 20 years different certification models have established themselves in the service sector in which organizational tools as listed in Fig. 1 are indirectly or directly required. These are partly cross-sector models, just like the DIN EN ISO 9001:2008[1] or even sector specific such as the VDR[2] – Certified Busi-

Fig. 4. The Qnigge® pyramid of organization

Source: Qnigge® GmbH – Freude an Qualität

---

[1] DIN EN ISO 9001:2008 is an international quality standard.
[2] The German Business Travel Asociation.

ness/Conference Hotels or the initiative "Service Quality Germany" (Service-qualität Deutschland), which approaches service providers in the tourism sector. All certification models have one thing in common, they ensure product and service quality and want to offer support so that a company has a good internal organization.

Company values and guidelines form the basis of the pyramid. Both also form the basis for corporate culture.

8   Company Values Derive from Qnigge®-Values as an Example

In his bestseller "Good to great ... why some companies make the leap ... and others don't" the American author Jim Collins impressively describes results from a long-term study. He compares eleven American public companies, which have successfully established themselves in the market for over 30 years. Finding out characteristics which these companies had in common was one of his goals. One of the success factors was company values, which were not only brought onto paper, but were fulfilled by actively living them.

This in turn has inspired us to think about our own company values, to set an example for other companies to find their corporate values as well and live these.

Fig. 5. Qnigge®-values

With this we managed to find a German term for each letter of our company name, which is important internally towards our employees and externally towards our clients and business partners.

- Q – Qualität (Quality)

  It's our aim to constantly reach or even surpass requirements and expectations of our clients in terms of quality. If we manage to do this and keep on inspiring our clients, then we are heading in the right direction.

- N – Nachhaltigkeit (Sustainability)

  We developed a pragmatic solution, which provides us with results and is sustainable, by collaborating with our clients, colleagues and employees. Furthermore are well aware of our responsibilities with regards to the environment and consider these for our trips. Our aim is to choose the most environmental friendly way of taking us to the desired place at the desired time. During 2009 and 2010 we can claim that we managed to save 10.000 liters of diesel by using public means of transport for business trips.

- I – Initiative (Initiative)

  The current status can always be improved. This requires curiosity, interest and initiative. This is the only way of further developing ourselves and by doing so for our clients.

- G – Gewinn (Profit)

  Being a business partner we equally value our customer relationships. All involved should profit from the collaboration and attain an additional value. Additionally being a business we must achieve profits to be able to be present in the market long-term.

- G – Glaubwürdigkeit (Credibility)

  We act according to our system of values. We make use of the contents, which we pass on during seminars and workshops, within our own surroundings.

- E – Emotionen (Emotions)

  Living, working, learning, quality – this all requires emotions. They are the basic elements, which make us human and help us passionately pursue our and our client's goals on a daily basis.

Five years ago I wouldn't have dared to publically write about or even demand the value "emotions". It is only personal development and observing our clients that doesn't leave me with any other conclusion: "Emotions are needed, particularly in the business world".

Another Example from Practice

In collaboration with ATLANTIC Hotels in Bremen the Qnigge® consulting philosophy's main modules were implemented. The quality management system of

ATLANTIC Hotels uses many of the recommendations given by the Qnigge® pyramid of organization. "Some things must be standardized if you want to get noticed", says Markus Griesenbeck, CEO of ATLANTIC Hotels. "The system is designed so that we can also attain an ISO-certification in one or two years." (Markus Griesenbeck, Allgemeinen Hotel und Gastronomie-Zeitung, Issue 2010/42, p. 9)

ATLANTIC Hotels also developed a set system of values with management guidelines. In 2009 the ATLANTIC Hotels Bremen with its five hotels and approx. Four hundred employees were facing their biggest challenge in the company's history. Three new hotels from the premium segment were integrated into the company from different locations. The number of managers and employees abruptly doubled. To master the growth challenge an extensive quality management project was launched in 2009, which aimed to include all eight hotels and its head office.

It was important to the management that directors, head office managers as well as hotel heads of department could incorporate their own ideas to develop a sustainable value and management system. Next to the vision, mission, company goals and values management guidelines were to be laid down to obtain a common guideline in the future. This is how the five company values were formed as well as a small crocodile called "QRGOL", a mascot which represents the company values.

Q ualität (Quality)
R espekt (Respect)
O ptimismus (Optimism)
G laubwürdigkeit (Credibility)
L oyalität (Loyalty)

Fig. 6. QROGL® – mascot of ATLANTIC Hotels Bremen
Source: ATLANTIC Hotels Bremen

The special challenge for the company was to harmonize the company's values with daily actions of managers and employees towards clients and colleagues. As we now know thanks to extensive surveys people's behaviour is determined by personal values and life motives (cf. Brand/Ion 2009, pp 18–20). That is why it is even more important that company managers are aware of this fact and act accordingly, to be successful in the long-term, creditable and likable.

9 How Well Do Employees Know Their Company's Values?

Thanks to a further survey and by using "Q[kju:]-Interactive" we were able to find out how well participants know their company's values. One hundred and ten out of 160 participants answered.

The question was: "Do you know your company values?" with different answers to choose from.

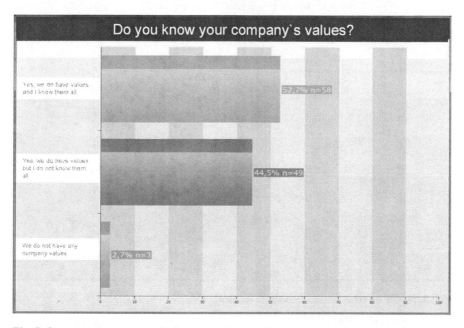

Fig. 7. Survey on awareness of company values

Source: Codiplan GmbH

If 50% of employees don't know all of the company values, then the question arises, how is it possible to integrate the defined aspiration into the company's quality of service, so that it noticeable to clients. The results evidently show that there is room for improvement as far as leadership is concerned.

Let's assume that quality is important to all successful companies and is embedded into the company's values. Then the question arises, who can explain what quality is?

10   What Do You Know About the Term Quality?

Using "Q[kju:]-Interactive" we are able to carry out further surveys asking participants how they understand the term quality.

The first question was: "What do you associate with *quality*?"

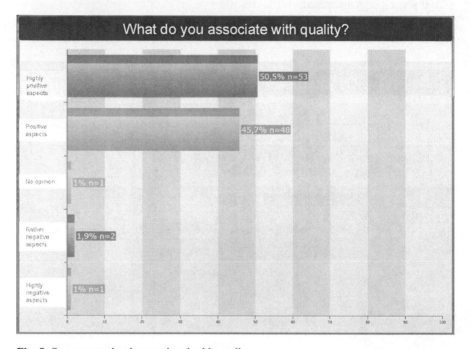

Fig. 8. Survey on what is associated with quality

Source: Codiplan GmbH

It is hardly surprising that over 96% of participants asked had a positive association. However if the term is translated literally, then another conclusion has to be made.

The next survey asked participants, what the German word "Qualität" (quality) stands for, because the term does originate from Latin after all (qualis, lat.). One hundred and forty-two out of 160 participants responded to the question.

Only 15% of participants answered correctly and chose "attribute". Quality is indeed a neutral term, which only turns into a value when considering the observer's point of view and his expectations. This in turn has drastic consequences for all service providers, because a service cannot be created through the perspective of the service provider but rather through the perspective of the customer und this unfortunately isn't always the case.

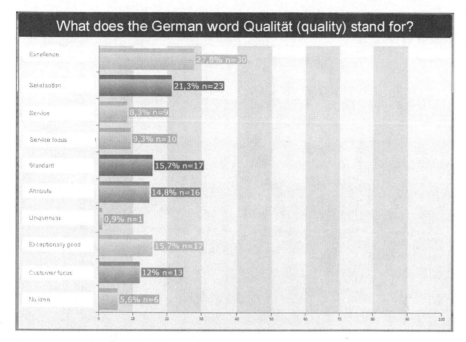

Fig. 9. Asking for a translation of the term quality

Source: Codiplan GmbH

When you look at some of the coffee breaks and their arrangements that go on during an event, then you have to ask yourself how it is possible that coffee cups and spoons are found in different places.

I as a conference guest expect that both of these items are found together at the buffet and in ample amounts. It is also nice if the coffee pot doesn't drip and tea isn't served from a tea pot previously used to serve strong coffee. These are matters that are taken for granted and the caterer is well advised not to take them lightly.

11 Evaluating the Quality of an Event

We were also interested in finding out whether single trades of an event had an influence on the participant's overall assessment of the quality of an event.

Once more we were able to determine the front runners within a few seconds using "Q[kju:]-Interactive".

The first question was: "Which is the most important service provider in your opinion in terms of having influence on the quality of an event?"

The occupational group event planners and PCOs managed the highest score at 34%. This value speaks for itself and when considering the event as a synthesis of

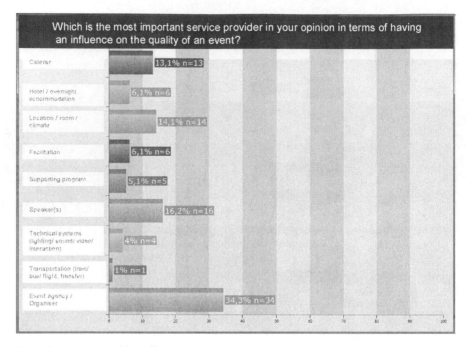

Fig. 10. Survey on which influence service providers have on the quality of an event

Source: Codiplan GmbH

art, then each and every service provider carries a high level of responsibility in terms of the quality of an event as a whole, whereby speakers (16%), the location (14%) and catering (13%) still attained the highest values.

The other results confirm the philosophy of the agency CODIPLAN Gesellschaft für Concept, Planung und Dialog mbH, as many parts make up the whole and this is reflected in the service offer of the agency. Establishing a brand is not just finding the appropriate way of passing on messages, but also developing contents competently. This "one-stop shop" concept holds the crucial advantage of optimally harmonizing form and content, strong synergy effect being created and that messages can achieve the most powerful impact.

Using this philosophy it is possible to eradicate quality deficiencies before they even appear, because the question asked about quality deficiencies produced the following picture.

The question was: "In which areas do you see the highest quality deficiencies in the entire event process?" Up to three different answers were possible.

The largest quality problems are obviously seen in agreements among trades such as catering, post-service/checking customer satisfaction and speakers/contents. These are clear messages to agencies, event planners and service providers to work on the elements of the event management process. Interestingly customer orientation can be found midfield in the ranking and isn't perceived as such a problem.

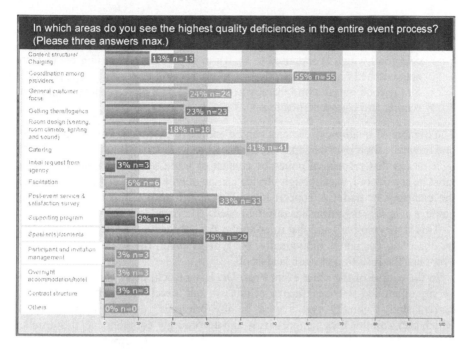

In which areas do you see the highest quality deficiencies in the entire event process? (Please three answers max.)

Content structure/ Charging	13% n=13
Coordination among providers	55% n=55
General customer focus	24% n=24
Getting there/logistics	23% n=23
Room design (seating, room climate, lighting and sound)	18% n=18
Catering	41% n=41
Initial request from agency	3% n=3
Facilitation	6% n=6
Post-event service & satisfaction survey	33% n=33
Supporting program	9% n=9
Speaker(s)/contents	29% n=29
Participant and invitation management	3% n=3
Overnight accommodation/hotel	3% n=3
Contract structure	3% n=3
Others	0% n=0

Fig. 11. Survey on quality deficiencies sorted by service providers

Source: Codiplan GmbH

At the same time the message is that service providers should not just think about how they can fulfill demands and expectations as a whole, but also make it more noticeable and exciting for participants and organizers through special service attributes.

The simple question is: "How can that what we do become better/customer friendlier/cheaper/more profitable/nicer …? The list of attributes can go on endlessly.

Some examples from practice from a speaker's or a conference participant's point of view:

1. Seating and conference tables.

   Again and again it reoccurs during meetings that you are seated at narrow conference tables, so that there isn't enough room for your own documents/laptop/glasses. Tables are also often set up so close to one another that participants sitting at the table in front of you knock over glasses and drinks on your own table with their jackets.

2. Quality of slides

   Some speakers provide slides with small writing, packed with lots of numbers and unreadable contents which is completely unreasonable towards participants. How is the message supposed to come across?

3. Buffets

Yet another example from the gastronomy sector: Who has never been served cold food which is supposed to be warm at a buffet? Warming plates were not switched on or sufficiently turned up, serving cutlery was inappropriate for the food on offer and thereby putting food on a plate became a time consuming and almost acrobatic act.

The number of quality deficiencies can be endlessly extended. Presumably everyone has made their own personal experiences.

It means that respective points of contact of trade service chains of an event must be considered, even though service chains of an event do differ depending on the company or trade. A concert has different demands compared to a football game, a congress organizer has different needs compared to a seminar provider.

This is why no hard and fast rules apply; the only way of achieving set goals is by observing customers individually, the type of event, the respective aims of an event and the resulting process landscape.

In the course of the same event Prof. Dr. Hans Rück, dean of the Worms University of applied sciences, faculty of tourism and travel management presented a current study "Quality events need quality briefings" during his presentation, which backs up this thesis impressively and accordingly complements this article.

12 Defining Chains of Service and Processes

The concept of "Initiative Servicequalität Deutschland" (Initiative Quality of Service Germany), which specially addresses touristic service providers with its three steps of development, also offers useful suggestions in the German-speaking world. The aim among other things is to describe service chains and processes and to improve these using selective measures on a regular basis. (www.servicequalitaet-deutschland.de)

Fig. 12. Example of a service chain

Source: Qnigge® GmbH

Each process step "involves" a series of individual processes that can be structured very differently depending on the company. Here each company is able to define its differences and stand out compared to its competition. The question is: "What can we do to thrill our customers?", what provides the "certain extra?", "what are the „magic moments'?". Often these are only little details.

Which "magic moments" are effective in the end may differ for each project and customer. There are also companies which have managed to standardize the little details. A short time ago I heard about an hotelier who paints a small smiley on all of his breakfast eggs, every morning, every day, every egg … That is a standardized process, certainly a small detail, which I have yet to see in any other hotel. Creativity knows no boundaries, provided that the standard processes work. What use is a smiley if the buffet is otherwise set up carelessly and the serving cutlery is difficult to use, of what use is the best lighting technology used to illuminate a conference room when the seats are uncomfortable and your backs starts to ache after a while?

13 Documenting Organizational Structure and Service Chains

Using organigrams, job descriptions, check-lists and process documentations provide the basis for being able to constantly implement processes, including the small details and also train new employees towards this. Thereby quality is not a random product but the result of a planned process orientated to values of a service chain.

In the meantime brilliant online guide tools, which make it easy to document service chains are available (for example www.orgavision.de, www.wissintra.de, www.bitqms.de).

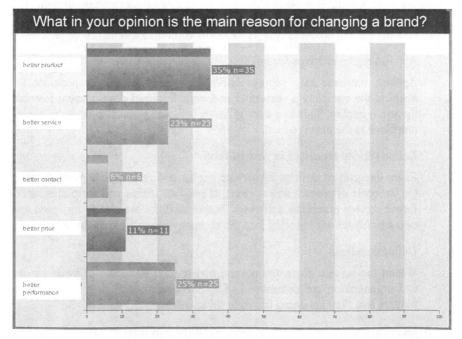

Fig. 13. Reasons for changing a brand

Source: Codiplan GmbH

14 Why Do Customers Change Their Service Provider?

The dialog finished with this sensitive matter. Using "Q[kju:]-Interactive" we were once again able to display the participant's opinions within a few seconds.

The concrete question was: "In your opinion what is the main reason for changing a brand?"

It is impressive to see that at the end of the day the price evidently plays a tangential role although of course every organizer is generally price sensitive.

Ultimately however it is important that the product, service and the entire service concept are harmonious and the targets of an event are achieved. What use is it if an organizer sends his employees to an event for lots of money and doesn't benefit in return.

15 Six Steps to Unmistakable Quality

Every company from the MICE industry, regardless of size and number of employees can use the following steps as a guideline to develop a quality organization.

1. Set fixed values

   Set your company's standards. Which values are important to you and should be a part of your service concept and get noticed by your customers, co-workers and business partners. Quality of course should not be lacking.

2. Appoint representatives for quality

   Appoint someone as a representative for quality who is responsible for aligning the company's personnel and organizational development towards the value quality. This is a very difficult task and should be done by someone close to the management.

3. Define service processes in written form

   Every department defines its service chains and processes in written form. On the basis of process descriptions, check-lists, forms and templates an efficient service process is achieved. Responsibilities are clearly assigned and new employees are inducted into the company using these documents.

4. Define standards

   Within the service chain the normal and special standards of your service and organization are determined. Thereby you manage to ensure all employees that they are doing the right things properly.

5. Defining unexpected WOW-effects, so called magic moments

   Always reinvent yourself and define unusual standards which make a difference.

6. Get feedback

As we can never be sure whether we are really satisfying our customers you make sure that customer and employee opinions are constantly integrated into the internal renewal process by using feedback.

16 Conclusion

Quality in the service sector nowadays must not be a random product, a result of individual services. It rather needs service processes, which are orientated towards the aims of different organizers and types of events, documented and implemented systematically.

Points of contact between service providers must be tuned and always checked systematically on a regular basis.

A shortage of skilled staff will also make it increasingly important for all parties involved in a company from the MICE industry to meet the demands placed on quality on a customer-focused basis in line with their corporate value. This needs a management system based on transparency and values. Own training programs help to ensure that also new and semi-skilled co-workers are quickly ready to work at a good quality level.

It is about a dynamic process which continuously registers changing customer demands and expectations and integrates these into the internal renewal process. Processes and employee behavior are systematically developed.

With this in mind enjoy quality with Qnigge® in the MICE industry.

References

Bruhn, Qualitätsmanagement für Dienstleistungen, 8th edition, Springer 2011
Collins, Der Weg zu den Besten, 8th edition, DTV 2008
Ion/Brand, Motivorientiertes Führen. Leadership using the 16 life motives according to Prof. Steven Reiss, 2009
Ueding (ed.), Adolph Freiherr von Knigge über den Umgang mit Menschen, Insel Verlag 1977

Quality Events Need Quality Briefings – Professional Communications Briefing as a Key Factor for Creating Successful Events

Hans Rück

1 Introduction

1.1 The "Trend Towards Events"

Events[1] are becoming more and more important – in tourism, and beyond.

Events are an essential reason to travel and a pivotal criterion when it comes to selecting a holiday destination. Touristic destinations, in turn, use events increasingly often as an instrument of differentiation from their competitors, which is shown for example by the continuous growth of festival and musical productions in the recent years. Similarly, business trips are initiated by events like congresses and seminars. The hotel industry participates by offering accompanying overnight hospitality services, the conference hotel business more so by offering meeting rooms and related services which are provided by their banquet departments. Even cruise lines and theme park operators are placing more and more emphasis on events to fill their ever growing capacities.

To use Horst Opaschowski's famous word, a "race of adventure worlds" is taking place in the modern leisure society, driven by a seemingly unquenchable

---

[1] Originally, the term *events* is used to denote special (or unique, spectacular, unforgettable) functions. However, this popular understanding is academically inappropriate because the defining characteristic "special" is ambiguous and not objective. Therefore, the term *events* is used here in a broader sense comprising organised gatherings of all kind, whereby the focus is laid on commercial events. It ought to be mentioned that this wider definition does not match the popular, but misleading acronym "MICE" which is supposed to represent and explain all events as the summation of meetings, incentives, conventions, and exhibitions, but in fact (and at best) comprises only "business events". Events in the sense of our wider definition are characterized by four features: they follow a certain aim and are not coincidental; they are organised and staged; they are interactive as the event experience is created through the joint contributions of host (organiser) and guest (attendee); and they are multisensory as they appeal to all senses. (For a more detailed definition, types and specialties of events see Rück 2011.).

"thirst for adventure" and a growing search for collective experiences (Opaschowski 2000, pp. 8, 68), the latter being partly a counter-movement to threatening isolation in an age of electronic communications media.

Beyond tourism, events are steadily growing in importance as a marketing and communication instrument. Such "marketing events" are organised gatherings carried out by corporations (companies, associations, parties) themselves with the aim to meet their own marketing and communications objectives (Rück 2011). Apart from the above-mentioned general "trend towards events", the steady growth of marketing events is caused by the massive advertising overload in the classic media communication (Kroeber-Riel/Weinberg/Gröppel-Klein 2008), whose actual competitive position is constantly worsening due to high media costs and a continuously decreasing communication performance. However, marketing events elevate themselves in multiple ways from the classic "above-the-line communication" paradigm: They create one-to-one situations in communication (contrary to the classic one-to-many situation), they focus on dialogues (not on monologues), activate participants through multi-sensuality and "live" atmosphere, and so at the same time create an emotional added value for participants and an emotional differentiation for the advertised product. Marketing events serve the current trend towards adventure consumption by creating and staging adventure worlds for brands and products. As platforms for personal meetings and interactions they create social relationships and networks between organisers and participants and in this way help establish "brand communities" (Rück 2011).

1.2   Problem Outline

Against this background the event industry has seen a ten to fifteen year phase of growth by now which has sometimes been quite turbulent. As far as the German event industry is concerned, barely any structures could be developed which would reach a normal level of professionalism.

Regulated job descriptions and generally accepted education courses do only in part exist (there are at least increasing efforts in this direction). Staff recruits therefore stem from a high percentage of lateral entrants coming from other professions, many without any previous knowledge in business administration, marketing and marketing communication. A lot of newcomers to the industry were and are of the opinion that having organized school parties would qualify them adequately to become an event manager. (This, by the way, does not seem to be any different in the Anglo-American world: Hoyle 2002, p.VI). It therefore should not be surprising that the image of event managers is still close to that of a party manager rather than that of a communication expert with conceptual capabilities -- which is exactly what an event manager should be.

In brief: Event managers still have a hard time to be taken seriously. There are no professional structures which would convey a clear, resilient idea of an event manager's capabilities and therefore provide some basic respect.

The procedures and processes in the event industry add to this picture, since they impede structured working. Systematic, long-term planning is still an exception, events are often set off in a "spur-of-the-moment" fashion, just like the notorious saying expresses, which in this industry has in the meantime become a "running gag" of typical aimlessness: "We have to do something again!"

This of course affects working procedures. Do something? Gladly! But for which target group? With which objectives? With which budget and personal resources? Questions like this get asked far too rarely. In other areas of the communication business – for example in classic advertisement – conceptual thinking is a daily routine and is demanded by service providers as a matter of course; briefings and briefing processes are conceptually applied accordingly.

Event managers are not expected to have the same conceptual abilities and the majority of event managers haven't contributed much towards changing this. Too many of them understand a briefing as an operative checklist to tick off: How many hotel rooms are required? Which artists may I book? Do we also need a flipchart? This of course has nothing in common with conceptual working, does however correspond to the traditional role of an event manager and is thereby exemplary for how much emancipation work alone has to be done so that a briefing is properly understood and that we still have a long way to go until event managers are taken seriously as communication experts.

In fact the briefing process is indeed very appropriate to exemplify the lacking professionalism in the event industry. Three figures emphasize this:

- On average only 46% of all German event agency briefings run flawlessly; this was the result of a survey performed by the Worms University of Applied Sciences, Rhineland-Palatinate, Germany. The survey results are later presented in detail.

- A survey of the German event agency MICE AG at the trade fair "STB Marketplace" in Munich on October 26, 2010 showed that 27% of the people asked agreed with the statement: "Event managers brief agencies insufficiently!"

- The same survey showed that 40% of the interviewees agreed with the statement: "Our events must become more professional!"

Considering these unambiguous votes we can register: Event briefings up to now deserve the mark "unsatisfactory". In the following we will take a closer look at the reasons and consequences for this problem.

It should be stated that for the methods applied the following portrayal as well as statistical data applied refers to the German event industry. However the findings are without doubt transferable as descriptions of similar problems in Anglo-American specialized literature show.

1.3   Objectives

The following article should:

- present particularities of event briefings and the event briefing process;

- point out event-specific problems which occur during briefings and clarify these on the basis of selected data of an empirical study carried out by the Worms University of Applied Sciences;

- show how briefings can be used as an instrument for event planning – especially for planning the "target effects" of an event – and illustrate this with the help of different event types;

- identify possibilities and starting points to improve cooperation between clients, event managers and agencies or service providers;

- demonstrate that briefings can serve as an excellent instrument to augment the importance of the event manager in the perception of his internal and external clients.

1.4   Analytic Procedure

In section 2 we shall first of all take a look at the origin and conceptual meaning of the term briefing, the importance of briefings for communication activities as well as the typical multi-stage procedures of the briefing process.

Then the ideal-typical procedures of an event will be presented and briefings will be allocated in this process.

In section 3 we shall present the results of an empirical study performed by the Worms University of Applied Sciences, which uncovers many faults in briefing structures in the event business.

In contrast to this section 4 will show which elements event communication briefings should ideally contain. To start off with, the basic elements of the event concept will be established. Subsequently, a flexible system for planning the "target effects" of an event will be introduced and demonstrated for different types of events.

This will be followed by taking another look at the changing role of event managers – and a look at the question what professional briefing structures can contribute towards meeting these changes.

This paper finishes off in section 5 with a summary of the most important results and a prognosis for the future of briefings in the event business.

2   Briefing: Term, Procedure, Meaning

2.1   Origin of the Term Briefing

Etymologically the term "briefing" derives from the Latin word "breve" (gen. brevis) which in the middle ages came to mean "letter, summary, short note", esp. from an authority, and yielded the modern, legal sense of "summary of the facts of a case" (1630s). In the military context, it took on the meaning of "to give instructions or information to sb.", which finally lead to the noun "briefing" in its contemporary sense, first attested 1910 and later on popularised by WWII pre-flight conferences (Harper 2011).

It is this military sense in which the term briefing is used today in the communication business. Let's take a look at how Back and Beuttler define this term for the communication business in their "Handbook Briefing" (2006, p. 10):

> *"During a briefing the initiator describes an objective for communication which the (internal or external) service provider has to solve by a given target date within the limits of a defined creative margin. To make this possible the briefing specifies the targets which are to be reached as precisely as possible, defines the creative margins and delivers information needed to understand the task in a very compact, well-structured form."*

We can take the following from this definition: A briefing formulates a task. It defines possible solutions by restricting financial resources and creative margins. At the same time the initiator has to enable the service provider to fulfil the task; this requires relevant information, which is to be supplied in a compact form.

Every briefing needs a re-briefing; this works as acknowledgement given by the service provider to the initiator so that he can be sure that the objective has been understood. Briefing and re-briefing form an inseparable unit, they are two sides of one coin.

2.2   Relevance of Briefings

Briefings gain importance wherever people are working towards a common goal. In an economy based on division of labour such collaborations are often organised in the form of service relationships. In the service context the briefing constitutes an effective and therefore widely used prescription for forming bite-size summaries of the cooperation essentials.

In most cases events are delivered as – internal or external – services which underlines the importance of briefings on the service side.

Characteristic for services is the collaboration of service provider and customer. (In the event business, the customer is the organiser or host of the event, not the guest or attendee.) The customer generally must participate in the production of the service and contribute at least partly to the final result (Maleri 1973); he is a "prosumer" – consumer and producer at the same time, that is (Toffler 1980). He has at least to inform the service provider in advance – either in the order details or in the form of a briefing –, how he would like the service to be carried out and provide necessary background information.

Therefore the quality of a briefing – its correctness, precision, and completeness – can decisively affect the service quality. No wonder the saying goes: Service providers can only be as good as their briefing. The more complicated the commissioned service is, the more important clear and binding agreements become. And events are without exception generally very complex services including many different trades where potentially numerous service providers get involved.

Additionally, briefings usually form the basis for procurement decisions. They prepare the ground for contractual relationships with service providers or specify these relationships. What is more, briefings can serve as legal proof in cases of dispute. For this reason briefings should always be given in written form (and in addition be explained verbally). This is why a re-briefing should come along with every briefing so that a uniform understanding of the task can be ensured.

Summarized we can establish the following; the importance of briefings for the success of an event can hardly be overestimated. A professional briefing provides all parties involved with certainty and ensures a confiding and positive atmosphere of cooperation. A professional briefing in the event business is definitely an *instrument of quality assurance.*

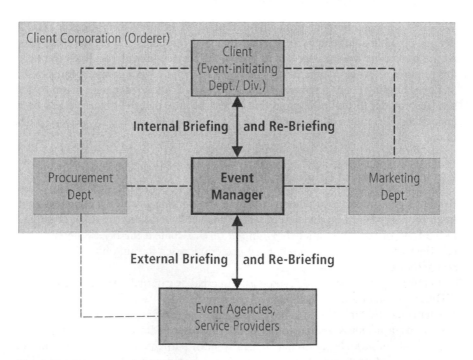

Fig. 1. Briefing as a multi-stage process

2.3   Briefing as a Multi-stage Process

Briefings are mainly described in literature as a single-step process between two parties (customer and service provider; see for example Back/Beuttler 2006, exception: Schmidbauer 2007). However, a single-step briefing process this does not reflect reality. In the communication business and especially in the event business briefings are typically multi-stage processes. When taking a closer look it's not just one briefing but a chain of briefings. Thereby internal and external briefings have to be differentiated (cf. fig. 1).

The start is usually made by the internal briefing which the event manager receives from his client (orderer), the event-initiating department or division. (We will keep the re-briefing for each stage in mind.) The event manager himself usually belongs to the firm's marketing or communication department and works as an internal service provider for each event-initiating department. This in turn is, just like the event manager, subject to the general purchasing guidelines of the company and is therefore partner of the procurement department.

The internal briefing is followed by the external briefing: Therein the event manager passes the objectives entirely or partly on to agencies and other service providers (catering, technical services, entertainment, security etc.).

Finally a third and yet again internal briefing takes place in the agencies (not shown in Figure 1), this time typically between the client manager and the creative director.

The event manager operates at the contact point between internal clients and external service providers. This is representative for his position and shapes his daily work which in many parts can be compared to a "transmitter" and "translator" function.

Now that we have worked ourselves through the briefing process as such we will see how and where briefings must be placed in the event process.

2.4   Phases of an Event

Overall an event can be split into five phases, two of which occur pre-event and two which occur post-event (cf. fig. 2):

- Phase (1) consists of the *event planning process*, resulting in the *event conception*.

This first phase is followed by the three *operative phases of an event:*

- The *Pre-event* phase (2) covers the event preparations, particularly the announcement/invitation and procurement processes.
- The *Main-event* phase (3) represents the actual event process "from set-up to clean-up".

- The *Post-event* phase (4) comprises the aftermath which normally includes activities like sending letters of thanks to the attendees and reports to the no-shows and to the media press.

- The event process is rounded off by another non-operative phase (5), the *event evaluation or event controlling*. However, it should be noted that controlling is a permanent task during the organisation of an event and should accompany all event phases (Luppold/Rück 2010: 260 f.).

Briefing – as it has been defined and is from here on understood – is without doubt a component of the planning phase of an event: It defines objectives and the permitted "solution space" in the form of target effects, target groups and target tools (the required or available resources), just to mention the most important elements.

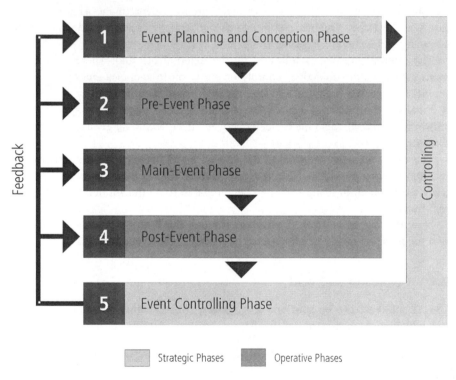

Fig. 2. Briefing as a multi-stage process

When taking a look at figure 2 it becomes obvious that the fixation of the event controlling guidelines is also part of the planning process and hence an element of the event concept. This is because event controlling necessarily must start off with the target effects and the corresponding "performance indicators" (for very important values also "key performance indicators") which are established in the planning phase: The measurement of the planned performance indicators constitute the central point of event controlling.

Thus, briefing should be understood as an pivotal element of the *event planning process*; not, as it is predominantly done today, as being mainly a part of the operative event process (i. e. the pre, main, post-event phases). This operative interpretation of a briefing corresponds to the more operative role many event managers still play today; but if event managers keep on treating briefings like checklists ("Do we have a flipchart there? Do we have beverages?" ...), they will definitely not meet up to the changing requirements of their field and, in the long run, damage the collective reputation of event managers beyond repair. (To avoid misunderstandings: Checklists as such have their purpose and a firm place in the operative event management; they are especially useful in process controlling in order to ensure error-free event processes. But that's it.)

To summarize at the half-way mark we can establish: Briefing is an instrument of event conception and not an operative checklist to be ticked off. We could also say: A good briefing is not about implementing – it is about thinking ahead! It is not about placing chairs – it is about placing messages!

3 Quality of Briefings in the Event Business: Results of an Empirical Study

3.1 Study Profile

In an empirical study carried out by the Worms University of Applied Sciences event agencies in Germany where asked to evaluate the quality of the briefings they get from their clients ("corporates"). For this purpose, 155 event agencies in Germany were questioned. The response rate amounted to 46.5%, the responding companies were made up of the following:

- 56% event agencies,
- 37% full-service advertising agencies,
- 7% professional congress organisers (PCOs).

Of all respondents 65% stated to be occupying upper or top positions (board of managing directors, managing director, director, divisional manager, company secretary).

3.2 Selected Results of the Study

The interviewees rated the importance of a quality briefing for the success of an event as "high". On average the sample of the survey saw a strong connection when asked to what extent the quality of briefings influences the success of an event (average rating 4.8 on a scale of 1 to 6; cf. fig. 3).

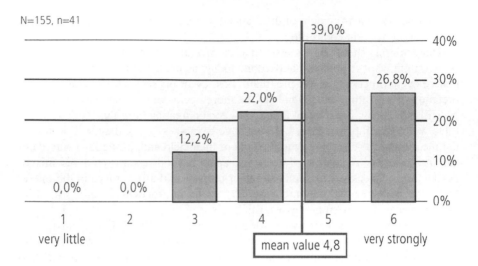

Fig. 3. How much does the briefing quality influence the success of an event?

Hence, it should be a cause for concern that every second briefing has shortcomings: The agencies answer the question "How many briefings out of ten run flawlessly" with an average rating of 4.6 – thus 46% (cf. fig. 4).

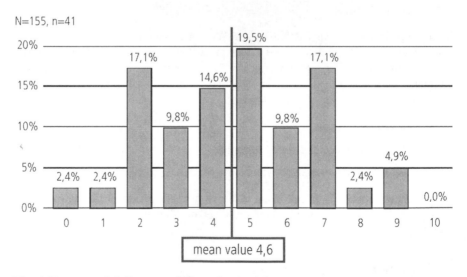

Fig. 4. How many briefings out of 10 run flawlessly?

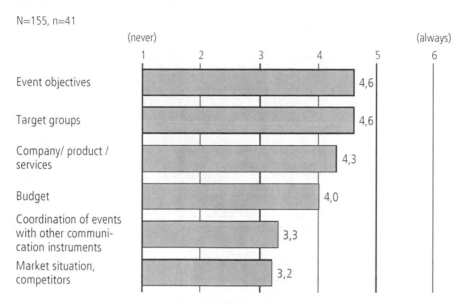

N=155, n=41

Fig. 5. Briefings contain information about …

This devastating result is obviously due to the fact that many briefings are missing important information. Information about event objectives and target groups is included often, but not always, as it should be (average rating 4.6 on a scale of 1 to 6; cf. fig. 5). Scarcely found is information on budget size (average rating 4.0), the company itself, its products and competitors (4.3 and 3.2 respectively) and information about the coordination of events with other communication instruments (3.3).

All of these points are mandatory components of any briefing. The fact that they are often missing is worrying. It probably isn't far off the mark to assume that the reason for this deficit is a mixture of lack of time and lack of quality awareness.

The list of the most frequent briefing problems is headed by "vague objectives" (average rating 4.4: cf. fig. 6). Formulating objectives seems to be one of the biggest problems in the event business: If any objectives are named in the briefing at all (cf. fig. 5), they are seldom precisely and clearly expressed.

Another main problem is the incomplete or missing delivery of information (average rating 4.4 and 3.7), whereby agencies don't exclude themselves from this criticism (average rating 3.2) – a fair gesture and also realistic.

Furthermore many interviewees complain that briefing processes often involve several people who hand out (different) briefings (average rating 3.7). The importance of this criticism is underlined by the fact that "constant contact persons" is stated as the most important criterion and "clear client responsibility" is the fifth most important criterion for achieving a good briefing (average rating 5.4 and 5.0; cf. fig. 7). Re-briefings in the implementation phase belongs to the same problem area (average rating 4.2; cf. fig. 8) – a bad habit which is unfortunately widespread in the bustling event industry.

N=155, n=41

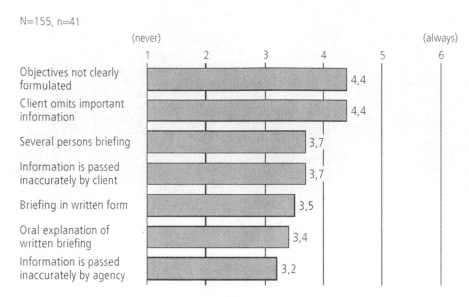

Fig. 6. Most frequent practical briefing problems

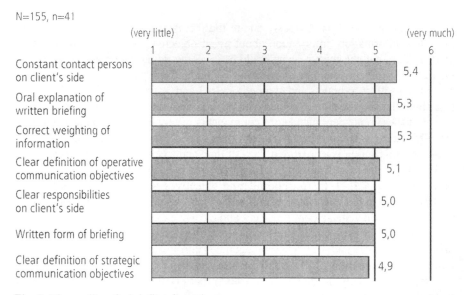

Fig. 7. The quality of a briefing depends on ...

Briefings are also often not given in written form and are not always explained verbally, as they should be (cf. fig. 2). The written form of briefings and its additional verbal explanation also count as one of the most important features of a good briefing. (cf. fig. 3).

Another significant criterion for high quality briefings is, according to the agencies, the definition of clear objectives, which includes operative event objectives as well as strategic marketing goals (average rating 5.1 and 4.9; cf. fig. 7).

However, according to the study results the client's company size does not play an important role for the quality of briefings (average rating 3.4; cf. fig. 8), which allows us to conclude that large firms do not per se have more professional briefing structures than smaller or medium-sized companies.

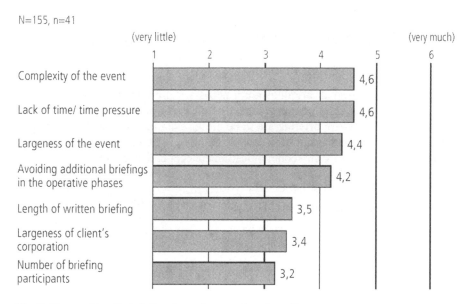

Fig. 8. The quality of a briefing depends on ... (continued)

The length of a briefing is also no yardstick for its quality (3.5) – this rating clearly shows that service providers do not like lengthy exegeses. Generally this may not be good news for the client's side – since summarizing information does, as is well known, cause more work than just plain listing does.

3.3   Interim Summary

The study results given above have confirmed our introductory thesis that the contemporary briefing processes in the event business urgently need more professional standards, especially concerning the following points:

- Briefings are often incomplete. Of all things conceptual and strategic information goes missing: about the event objectives, the overriding marketing objectives, the company, its products and services plus its competitive advantages.

- The available information leaves much to be desired. Especially the event objectives – if included at all – are often not clearly and precisely formulated.

- Briefings are often not made in writing – and even more often not explained orally.

- Contact persons on the client's side often change or their competences are not clear – even though constant contact persons are without doubt a main factor of success for an effective and efficient cooperation.

4   Briefing as an Instrument for Event Planning

4.1   Elements of Event Planning

As already demonstrated, briefings are a necessary preliminary stage for a communication concept. The planning aspect of an event and the resulting "should be"-contents of an event briefing are to be closer examined in the following, particularly as this aspect has not been considered often in literature so far.

The main question for any communication concept can be expressed in short:

- What is the target group (TG)?
- What are the target effects (TE)?
- Which target tools (TT) are to be used?

The three elements namely target groups, target effects and target tools are also key elements of any communication concept (cf. fig. 9).

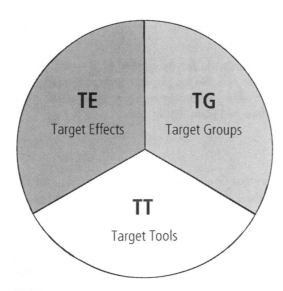

Fig. 9. Elements of event planning

At the beginning it had been already pointed out: Briefings in the event business nowadays are predominantly not of conceptual nature but focus on the operative event planning problems. In other words: The question of *target tools* is by far overrepresented in event briefings today. Compared to that the stronger conceptual elements "target groups" and especially "target effects" are more or less strongly neglected.

As far as we can see, the question of *target groups* is raised in most briefings in the event business, although not being sufficiently analysed and answered. As far as target groups are concerned, there do not arise any big differences between events and other communication instruments; therefore we will not elaborate this point here any further.

However, the field of *target effects* – as our study has proven – is the one which is neglected most strongly and often in event briefings. This is why we are going to pay the most attention to this point.

Which target effects do event professionals expect from their events today? Another survey carried out by Worms University of Applied Sciences from 2008 showed the following proportional distribution of planned effects (cf. fig. 10; the specific question was asked about marketing events).

Events are mostly used for image improvement (1st), furthermore for customer loyalty (2nd) and to increase brand recall (3rd). "Repositioning brands" touches the field of image management again (4th). Winning new customers (complementing

N=650, n=189; Collaborative Study of Worms University of Applied Sciences and MPI Germany

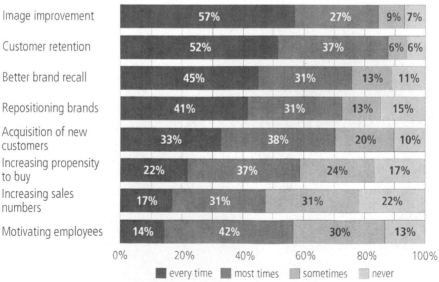

Fig. 10. Target effects of marketing events (actual distribution)

customer retention) (5th) and on "willingness to buy" is followed by "increase sales volume" and "motivate employees".

This overview gives an idea of the practical importance of specific target effects for events. However it does not provide any assistance to answer the crucial question: *Which target effects should be aimed for at which events?* This needs a system which allows a reliable allocation of certain target effects to certain types of event.

4.2   A Planning System for Target Effects of an Event

An effective system for planning target effects which can be used for briefings must be founded on a comprehensive and differentiated picture of the possible target effects of events. These can be divided into four types (Schäfer-Mehdi 2009; compare Rück 2006): Events can

- spark emotions,
- impart information,
- motivate to act, and
- initiate actions.

These four types of event target effects are shown in the following figure using different colours: Red stands for emotion; blue for information, yellow for motivation and green for action (cf. fig. 11).

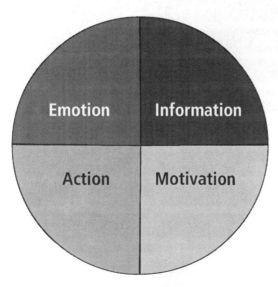

Fig. 11. Possible target effects of events

The relative weight of these four *effect components* varies depending on the type of event. This hypothesis is of critical importance as it allows to designate certain *combinations of target effects* to certain event types (Rück 2006). By way of example Figure L shows the *target effect mix* of four different event types. (cf. fig. 12):

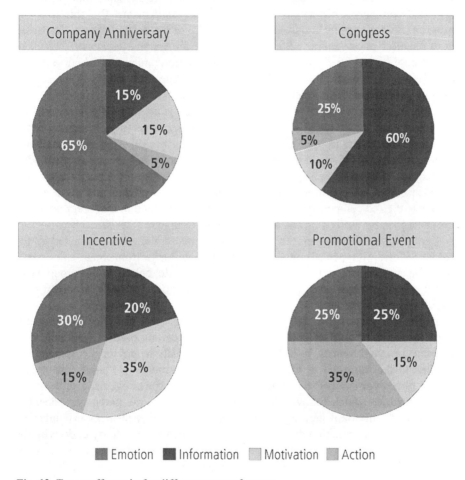

■ Emotion   ■ Information   ▨ Motivation   ▨ Action

Fig. 12. Target effect mix for different types of events

Some events specifically aim for *emotional effects*; here the colour red is predominant. Ceremonial acts, galas, anniversaries and also sponsoring events belong to this group. As an example, let us take a look at a company event celebrating its 100-year anniversary: Such a function primarily aims at strengthening the staff's sense of togetherness and to remind about the same values and conviction as well as about improving the company's image or reposition it in certain image components. (Image effects generally are counted among the emotional effects: Kroeber-

Riel/Weinberg/Gröppel-Klein 2008.) Presume, for example, the company is not regarded as a socially caring employer; then perhaps a historical review could enable to carve out the company's social commitment as a constant variable in its history in order to attain the desired image change. This would be the emotional component of the target effect mix *(red component)*. At the same time, information (the company's history) is passed on *(blue component)*. But this happens more or less incidentally and is not the focal point of the desired effects. The same goes for motivation *(yellow component)* and action *(green component)*, which in this example could be expressed in a lower resignation rate and/or labour turnover rate. However, it would be quite difficult to determine such effects as the sole or priority yardstick of such functions.

Other events focus on *cognitive effects;* this type is consequently dominated by the colour blue. Education or science events (congresses, meetings, symposia, seminars and workshops) as well as pure information events (like general assemblies, press conferences) belong to this group. Of course congresses are not only attended for reasons of knowledge acquisition; at the same time they are a platform for personal meetings, an opportunity for socializing and networking, and these effects belong to the red component. But they normally do not stand in the foreground, and the same goes for motivation and action which basically are side-effects here.

Further types of events aim for *motivational effects;* these are dominated by the colour yellow. All types of motivational events, e. g. kick-off-meetings, sales representative conferences, staff and distributor incentives etc. belong to this group. Generally speaking these work towards changing behavioural attitude, whereby concrete aims can vary greatly. Let us use "fam trips" (familiarization trips) as an example: The purpose of "fam trips" is to familiarize potential clients or procurement managers of tourism or event services with a certain product, generally destinations or event venues, to influence their propensity to buy or recommend the product to the customer at the counter (yellow component). Direct behavioural effects (green component) are important, too, for this type of events, e. g. the number of inquiries which follows such an event; however, the green component is only second important here as many other factors and influences can interfere before a certain destination or venue is booked which have nothing to do with the event itself. Emotions (red component) are pivotal for motivational events, since motivation cannot be achieved without enthusiasm. Lastly, information belong to the mix as well, since recommendations or procurement decisions have to be grounded on a minimum of information. However, the cognitive component is of comparatively little importance here; in the center of the mix are motivation (i. e. the propensity to act), action itself, and emotions.

Finally there are events which are supposed to trigger *concrete actions;* here the colour green dominates. Typically belonging to this group are all types of sales promotion events such as product presentations, point-of-sale events and road shows. The priority target effect is of course direct action, such as purchasing, followed by motivational and emotional effects (enthusiasm for the respective product), and information (e. g. about product characteristics).

Generally it can be said that emotional and cognitive conviction form the basis for the motivation to act in a certain way and – when the motivation is high enough and the situation is right – action itself.

We can see that no event is like the other: All events in principle bring forth the same types of effects; however, the target mix is always different.

This simple basic idea leads to a flexible and universally applicable planning system for target effects. The first step involves specifying the event type, the second step the mix of target effects and its relative weight. This configuration is, at the same time, the core of any communication briefing.

By deciding on the target effects of an event the event controlling is predetermined as well:

To each target effect belong certain performance indicators and measuring instruments or methods. The sequence "target effect – performance indicator – measuring instrument" forms a "triple jump" which allows us to measure the success of an event in a completely logical and easily understandable way (cf. fig. 13).

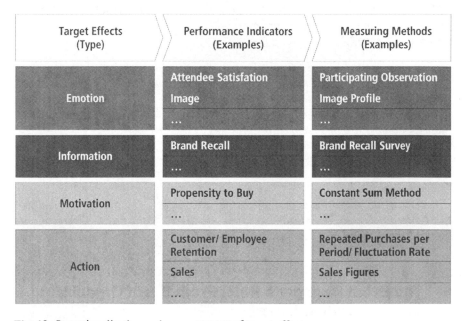

Fig. 13. Operationalisation and measurement of target effects

The topic of event controlling can only be briefly touched in this article, which is why we will restrict ourselves to a few explanations of the figure 13 above. If, for example, "attendee satisfaction" has been defined as one of the emotional ("red") target effects: This could be measured either by verbal interviews or by participatory observation; the corresponding performance indicators could be the answers on an open interrogation (interview) or a "facial affect scoring" (observation). Or let us assume that "improving customer retention" has been defined as one target effect in the (green) action field: This could be for instance operationalised by the number of repeated purchases in the month after the event and be measured by the sales statistics. The examples could simply go on using this scheme.

Conclusion: Event controlling proves to be a direct effluence of the event planning. During the briefing process the measurement of event success is determined together with the target effects.

4.3   Structure of an Ideal Communication Briefing for Events

Which components should be included in a communication briefing? The following table shows the basic structure of such a briefing (details are implied, but not discussed at length).The focus should definitely be on the section "event planning".

1. Project and Project Team
What is the project about (communication object) and what is its task? Pro-ject description, job data, contact persons, contact information, functions, responsibilities, competences.

2. Planning
2.a) What are the strived for target effects? How should these be evaluated before/during/after the event?
2.b) Which target groups are to be addressed? Which information is given about market, competitive and competitor advantages?
2.c) Which target tools should be used? Budget, deadlines, destination and location, standards given by superordinate strategies, expert design wishes etc.

3. Organisation
Checklists for disposition and operative execution of logistics (arrival/departure, hotel etc.) and technical crew (location, equipment, technical expertise, catering etc.)

5   The Event Manager's Role: From Event Organizer to Communication Expert

The introduction has already shown: The financial weight of events in the corporate communication mix has constantly increased in the last ten to fifteen years and it is expected that it will further increase, albeit at a lower rate, in the years to come.

This development enforces a professionalisation of communication management which goes along with a profound change in the role of an event manager. With regard to the pharmaceutical market the international management consultancy Deloitte (2011:8) summarized the changes as follows:

> *"In most of the companies we have interviewed the role of an event manager is changing from a mere operative function to being a partner and consultant of the business divisions. The logistical execution of events and many supporting activities are being detached from the event manager's area of responsibility and passed on to external service providers. Therefore more conceptual tasks will in future be in the focus of the event managers' work: Experience and expertise in planning and development of events, discussing the target group focus and selection with marketing and sales, foster the introduction of new media etc. To sum it up: increasing the value of events for the company by using a wide spectrum of activities and by improving the management of events."*

Figure 1 has shown that event managers usually move within a polygon of partners and have a kind of mediator function in this network of relationships. This is where in future the event manager will have to position and prove himself as an expert for "live communication" and as a conceptual mastermind to internal clients as well as external creative partners. Event organisation and logistics which have been the traditional main feature of an event manager's activities is now by contrast losing importance and is outsourced increasingly often to specialized service providers.

The event manager's transition from an event organiser to a communication expert corresponds to the transition of briefings from a checklist to a conceptual framework. Professionally managed communication briefings can prove to be a highly effective instrument for improving the reputation and position of an event manager as a communication expert and ensure him sovereignty over the event planning process. In briefing talks the advantage is usually on the side of those who have a clear understanding of the problem structure and the fitting solutions. This structure is clearly mapped out in the communication briefing shown in section 4.3, whilst the client often does not have a clear idea of the target effects, target groups, and target tools which make up a successful event concept.

6   Conclusion

The role of the event manager is shifting more and more from organisational to conceptual tasks, and from the position of an operative instruction recipient to that of a strategic business partner and consultant of the company divisions who is at eye level with his clients.

In this article it could be shown that the communication briefing is a pivotal element in this change of role since it forms the conceptual framework for event planning and is, therefore, an indispensable element in the event planning process.

However, briefings are not used to their full conceptual capacity in event management today. Instead they are still frequently misinterpreted as a mere checklist for event organisation. What is more, an empirical study of the Worms University of Applied Sciences in the German events market has shown massive practical deficits in briefings and the related processes: Only half of the briefings can be regarded as satisfactory and complete; especially conceptual information is frequently missing, in particular when it comes to the goals and target effects of an event. In numerous cases, briefings are not given in written form and in even more cases they are not verbally explained, as it should be. And Briefings are often conducted by more than one person while the capacity and authority of the partners involved remains unclear.

From these empirical insights we can draw the conclusion that there is still much room for improvement regarding briefings and briefing processes in the event business.

Professional structures would require that briefing is first of all discovered and then utilised to its full potential as an instrument for event planning – especially for the "target effects" of an event. In this article we have introduced a sensible, practical easy-to-use model for planning event effects and illustrated its mechanism using the example of various event types.

By this means, we have demonstrated that briefings can serve as an excellent instrument to augment the importance of the event manager in the perception of his internal and external clients, and how professional briefing processes can help the event manager to live up to his new task as an expert in "live communication". Hopefully, event managers will gradually adopt their new role; because events as a communication instrument have in the meantime become far too important, that unprofessional structures could be accepted in the long term.

References

Back, Louis/Beuttler, Stefan (2006): Handbuch Briefing: Effiziente Kommunikation zwischen Auftraggeber und Dienstleister, 2nd ed., Stuttgart

Deloitte Consulting (2011): Effizienz im Pharma-Marketing: Veranstaltungsmanagement zwischen Kostendruck und Marketingerfolg. Berlin

Harper, Douglas (ed.) (2011): Online Etymology Dictionary, link: http://www.etymonline.com/index.php?allowed_in_frame=0&search=briefing&searchmode=none, October 9th, 2011

Hoyle, Leonard (2002): Event Marketing: How to Successfully Promote Events, Festivals, Conventions, and Expositions. New York

Kroeber-Riel, Werner/Weinberg, Peter/Gröppel-Klein, Andrea (2008): Konsumentenverhalten. 9th rev. ed., Munich

Luppold, Stefan/Rück, Hans (2010): Event Controlling and Performance Measurement. In: Buck, Martin/Conrady, Roland (eds.): Trends and Issues in Global Tourism 2010. Berlin, pp. 253–277

Maleri, Rudolf (1973): Grundzüge der Dienstleistungsproduktion. 1st ed., Berlin.

Opaschowski, Horst W. (2000): Kathedralen des 21. Jahrhunderts: Erlebniswelten im Zeitalter der Eventkultur. B.A.T. Freizeit-Forschungs-Institut GmbH, Hamburg.

Rück, Hans (2006): Wirkungen von Marketing-Events steuern und messen: Event-Navigator hilft als neues Instrument. In: Otto-Rieke, Gerd (Hrsg.): Modernes Geschäftsreisemanagement. VDR Jahrbuch, vol. 8, Munich, pp. 150–155
Rück, Hans (2011): Events. Gabler Wirtschaftslexikon Online, Sachgebiet Tourismus, http://wirtschaftslexikon.gabler.de/Definition/event.html
Schäfer-Mehdi, Stephan (2009): Event-Marketing: Kommunikationsstrategie; Konzeption und Umsetzung; Dramaturgie und Inszenierung. 3rd ed., Berlin
Schmidbauer, Klaus (2007): Professionelles Briefing: Marketing und Kommunikation mit Substanz: Damit aus Aufgaben schlagkräftige Konzepte werden. 1st ed., Göttingen
Toffler, Alvin (1980): Die Zukunftschance. München

Corporate Social Responsibility in the Tourism and Travel Industry

CSR and Sustainability in the Global Tourism Sector – Best Practice Initiatives from the Public and Private Sector

Taleb Rifai

As the world confronts economic uncertainty and environmental challenges, a consensus is emerging among governments, businesses and civil society that "business-as-usual" is no longer an option. Instead a new, more sustainable, global economic model is needed; one that can integrate social and environmental concerns in all that we do. This new model has been identified as the Green Economy and is rapidly gaining ground as our most viable option for a sustainable future.

In the move towards a Green Economy that can deliver on the three pillars of sustainability – economic development, social equity and environmental protection – travel and tourism, one of the world's largest and fastest growing economic sectors, will be key. For its full potential as a driver of sustainable growth to be realized, however, it will need concerted action from all involved.

1 Sustainability in the Tourism Sector

In 1950 there were just 50 million international tourists in the world. By 2010, international tourist arrivals had reached 940 million and are expected to reach 1.6 billion by 2020. UNWTO estimates domestic tourism to be around four times the size of international tourism; amounting to nearly four billion domestic tourists in 2010.

This growth in arrivals has been accompanied by a similarly impressive growth in international tourism receipts; reaching US$ 919 billion in 2010. Tourism today represents 5% of the world's GDP; one in twelve of its jobs; and 30% of its exports in services.

The sector has become one of the driving forces of global employment, economic security and social well-being of the 21st century. Today, in the face of a multitude of global challenges, from persistently high poverty rates to job shortages, the world needs the energy that tourism generates.

Yet together with the opportunities of job creation and development generated by tourism, come the pressing challenges of sustainability and responsibility.

2   The Move Towards Sustainability

It is well-documented that mismanaged tourism development can have detrimental impacts on the very environment on which it depends for its sustained success. The risks of over development and over exploitation of natural resources are real and cannot be ignored. The positive news, however, is that sustainable tourism is rapidly assuming greater importance among all involved in tourism and is now the rule, rather than the exception.

Sustainable tourism defined by UNWTO is one which "meets the needs of present tourists and host communities while protecting and enhancing opportunity for the future. It reflects not a trade-off between economic growth and the protection of natural and cultural environments, but the synergy between them".

This shift towards sustainable tourism, promoted and supported by UNWTO, has emerged alongside increasing concern among the international community that the very economic models that have allowed such an extraordinary growth over the past decades are placing our planet at risk.

Against this background, UNWTO actively promotes the Global Code of Ethics for Tourism – a set of guidelines and principles for all those taking part in the tourism sector – designed to minimize any negative impacts of tourism activity on destinations and host communities.

Over the last few years in particular – with financial, economic, food and energy crises placing into question the growth paradigms of the last decades – the idea of a new growth model, which of a "Green Economy", has emerged and moved firmly into the mainstream.

A Green Economy is one that results in "improved human well-being and reduced inequalities over the long term, while not exposing future generations to significant environmental risks and ecological scarcities" and represents a decisive step towards sustainability.

3   International Tourism in a Green Economy

International tourism has been identified within the Green Economy debate as one of ten sectors, alongside manufacturing or energy, which can lead the transformation to this new model.

According to the 2011 Green Economy Report, tourism is one of the most promising drivers of growth for the world economy and, with the appropriate investment, can continue to grow steadily over the coming decades, contributing to much-needed economic growth, employment and development while mitigating its environmental impacts.

The correct investment in green strategies would allow the sector to continue to expand steadily over the coming decades while ensuring significant environmental benefits such as reductions in water consumption, energy use and CO2 emissions.

With this investment, significant reductions in water consumption (18%), energy use (44%) and CO_2 emissions (52%) are possible, as compared with a "business-as-usual" scenario.

In addition, green tourism would stimulate job creation, especially in poorer communities, with increased local hiring and sourcing and a positive spill-over effect on other areas of the economy. The direct economic contribution of tourism to local communities would also be increased; maximizing the amount of tourist spending that is retained by the local economy.

It is clear that an investment in green tourism is an investment in sustainable global development. Investing in environmentally-friendly tourism can drive economic growth, lead to poverty reduction and job creation, while improving resource efficiency and minimizing environmental degradation.

4 Best Practices in the Public and Private Sector

Given tourism's sheer size and reach, even small changes towards greening can have significant impacts. But for tourism's full potential as a driver of a Green Economy to be realized, all those involved in tourism, from governments to businesses, will need to act. And act now.

4.1 The Public Sector

Governments have a key role to play in the move towards green tourism, namely through establishing sound regulatory frameworks, facilitating public investment and incentivizing private engagement.

It is up to governments to prevent unsustainable practices, create standards and provide incentives for green investment. Decreasing the cost of renewable energies should be pursued and adequate legislation on energy management and buildings performance is required.

Public financing is essential for jumpstarting the green economic transformation. Governments can facilitate the financial flow to the tourism sector by prioritizing investment and spending in areas that stimulate greening. Through public-private partnerships, governments can help to spread the costs and risks of large green tourism investments.

At the same time, government spending on public goods such as protected areas, water conservation, waste management, sanitation, public transport and renewable energy infrastructure can reduce the cost of green investments by the private sector in green tourism.

4.2 The Private Sector

The private sector has a key role to play in building a low-carbon, resource-efficient future through Corporate Social Responsibility.

More and more, tourism business leaders are aware that sustainability is no longer an option. More and more companies are investing in business models which preserve the environment, respect local communities and contribute to socio-economic development.

UNWTO research shows companies are particularly committed to projects related to the conservation of wildlife and ecosystems; the use of new technologies related to energy, water and recycling; education; and the development of local supply chains.

These projects should be applauded, promoted and expanded. Innovation should remain top of the agenda for the private sector, together with productivity improvement through efficient equipment use, savings from fossil fuel substitution and local and global carbon markets.

Steps towards more environmentally-friendly business strategies are not only the right moves ethically, they also make clear business sense. Companies are increasingly aware that green initiatives give them a competitive edge; build trust and brand loyalty; help them to retain customers, as well as recruit, keep and motivate employees; and result in reduced overall expenditure.

5   Global Initiatives

In addition to individual efforts from governments and businesses, it has become clear that global initiatives are required to address this global challenge and drive the sustainability agenda forward.

A clear example is the Global Sustainable Tourism Council, of which UNWTO is a founding partner, which is dedicated to promoting sustainable tourism practices around the world. Through a set of voluntary principles, the Global Sustainable Tourism Criteria, the Council works to expand understanding of and access to sustainable tourism practices.

The criteria address four key objectives: maximizing tourism's benefits for local communities; reducing tourism's negative effects on cultural heritage; reducing harm to local environments; and demonstrating effective sustainable management. Currently, over 70 organizations and companies are members of the Council and the Criteria are being implemented around the world by major tourism companies.

A second example is the Global Partnership for Sustainable Tourism (GPST), an initiative designed to connect all stakeholders working in sustainable tourism. Comprised of five UN organisations, governments, multilateral bodies, the private sector and non-governmental organisations, the GPST represents a common platform from which stakeholders can transfer experiences, record successful initiatives and replicate them to meet global needs.

Finally, Hotel Energy Solutions, a UNWTO-initiated project co-funded by the European Agency for Competitiveness and Innovation, and implemented in partnership with UNEP, the International Hotel & Restaurant Association (IH&RA),

the European Renewable Energy Council (EREC) and the French Environment and Energy Management Agency (ADEME), is an initiative aimed at increasing energy efficiency in European small and medium hotels by 20% and their use of renewable energies by 10%, demonstrating that economic growth and sustainability can, and should, go hand in hand. Its principal asset is a software – the Hotel Energy Solutions E-toolkit – which allows hoteliers to assess current energy use and decide on the most advantageous technology investment solutions. The E-toolkit is available free charge to all accommodation units registered with the project. While designed for European Union Member States in line with EU Energy Policies, the project is expected to be rolled-out globally over the coming years.

6   Sustainability in the Tourism Sector: A Collective Responsibility

Tourism is clearly a promising and dynamic economic sectors that can significantly contribute to sustainable socio-economic development while respecting the natural and cultural environments of host societies.

For tourism to play this role, however, a suitable policy framework is needed, plus a coordinated effort between public authorities and private stakeholders. Global initiatives must also be fully embraced. The sustainability challenge is a collective challenge and one we must all work together to overcome.

The CSR-System of Tourcert and the Effects of CSR Systems on the Company

Johannes Reißland and Petra Thomas

1   The Tourcert CSR Process

1.1   Introduction

The term corporate social responsibility, CSR for short, has gained a lot of importance in the past few years. As a result the international ISO 26000 standard, which came into effect in November 2010, was designed so that society acts responsible. Next to just considering environmental sustainability further issues such as human rights, working conditions, fair business practice, customer interests and developing the local community have also come to the forefront. The ISO 26000 is thereby supposed to be a guideline for companies and help them meet expectations when taking on social responsibility (ISO, 2011).

All basic ISO 26000 features were already included when Tourcert began to develop its CSR system in 2004. Reason for the development was the joint goal of the EED (Protestant Development Service), KATE (Centre for ecology & development) and forum anders reisen e.V. (enterprise association for sustainable tourism) of making sustainability in travel companies transparent and conclusive. For this purpose a CSR reporting system should be introduced. The criteria catalogue of forum anders reisen e.V., where criteria for environmental friendly and socially acceptable travel are stated, was used as a basis for defining criteria, which determine sustainability in tourism. The criteria are supposed to stimulate a sustainable development of tourism (forum anders reisen, 2004). Companies receive certificates from the non-profit certification society Tourcert GBR when they have successfully passed the CSR process.

The German Federal Environment Foundation (DBU) has financially supported the development and establishment of Tourcert's CSR process since 2008. Forum anders reisen e.V. obligated itself in 2008 that all tour operators who are members of forum anders reisen will have to have begun or completed the CSR process by the end of 2011. This is supposed to take the association a step closer in making sustainability, in which it has strongly believed in since its foundation in 1998, more transparent and conclusive for others.

By March 2011 51 travel operators had already been certified by forum anders reisen e.V., as well as three tour operators not belonging to the association.

1.2   Basics of CSR Reporting

In general the embracement of sustainability in companies statutes, goals and the company mission is desirable. When looking at contributions of companies towards sustainable development comprehensive attention should be paid on companies taking on more social and ecological responsibility in its core business than the law stipulates. It is not so much about single good deeds, but more about the stance and strategic orientation which is embedded into the company (Tourcert, 2011). Therefore significant feature of Tourcert's CSR process is concentrating on a company's core business.

The company screens its overall business activities by using given indicators. As a result from this a CRS report is composed following clearly defined rules. It is important that stated information is truthful, relates to the core business and is thereby significant. Furthermore it is described as clearly and understandably as possible, generally comprehensible and processed according to guidelines. Therefore it is comparable with other companies and conclusive. This standardization allows interested customers and other stakeholders to draw up comparisons between certified companies. CSR reports are read and checked by external assessors before the council for certification makes a decision on whether it should award the seal "CSR-Tourism-Certified".

Fig. 1. Requirements for CSR reporting

Source: Authors

1.3   Eight Steps to the Sustainability Report

The Tourcert system for creating a sustainability report consists of eight steps and starts with the management making a decision to the final external check by assessors.

CSR managers, who are taking charge of working on the companies process and are responsible for imparting the requisite expertise within the company.Therefore groups are trained in a course of an introductory workshop. In the further course product managers of tour operators are responsible for data collection. The management software "Avanti" is available to provide help for evaluating data. Once data have been analyzed, improvement potential is identified and

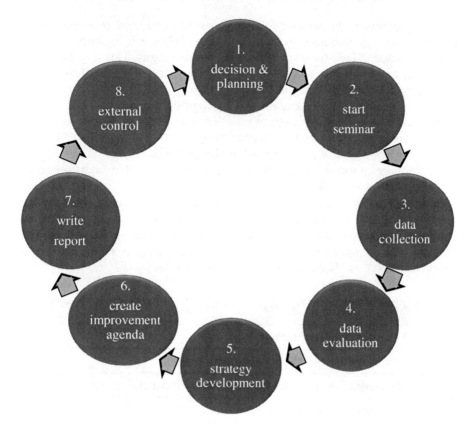

Fig. 2. Eight steps to the sustainability report

Source: Authors

strategies are developed on how this can be implemented into the business process. The improvement program is defined in the CSR report and thereby transparent to all. The seal is awarded after a successful external examination.

1.4   Key Indicators

Within the framework of the CSR process 219 indicators in total are taken into consideration, from which ten are defined as key indicators due to their significance. These should be published in the companies" sustainability report.

CO^2 emission, which mainly results from transportation to holiday destinations, is stated per guest/day. Based on the criteria catalogue of forum anders reisen there are three distance levels which are to be distinguished: Up to 700km, between 700 and 2000km and more than 2000km. Benchmarks can be created by forming averages and classifying using these levels. Exceeding a set amount leads to companies not being allowed to be certified.

Energy consumption per employee is generally considered relatively low as opposed to energy consumed by travellers. Nevertheless it is important to record this as well and to look for ways of bringing improvements. This launches a consciousness and awareness process in the company and it is gernerallyhelpful for embedding sustainability thoughts into the company.

Economic sustainability measures how much of the travel price remains in the destination. The information given to travellers is regarded as another significant aspect, as only an enlightened traveller will pay attention to sensitive and important aspects of sustainable travelling and implement these.

Furthermore customer satisfaction and a general successful business belongs to these key indicators, which are of importance for sustainable company management as well as for general quality management of tour operators. With the tour operator being able to influence a trip's entire value-adding chain as organizer and purchaser, evaluation functions also include examining whether the partners they do work with are sustainable in terms of their approach to business, the type of accommodation they choose and the tour-guides they deploy.

	Key Indicator
1	CO_2 emissions per guest/day
2	business ecology: CO_2 per employee
3	percentage of price, which flows into the destination
4	quality of customer information
5	index of consumer satisfaction (incl. return rate)
6	corporate culture: index of employee satisfaction
7	business success: cash flow and turnover
8	index of sustainability for business partner
9	index of sustainability for accommodations
10	index of sustainability for tour guides

Fig. 3. Key indicators of Tourcert CSR-system

Source: Tourcert, 2011

1.5   Certification

The CSR certification takes three steps. The sustainability report, which is commissioned by Tourcert, gets checked by an independent external assessor, who follows CSR reporting standards and guidelines during the first step. The check basically involves examination of documents and requests further information and certificates if necessary. The CSR assessors make sure that the report was written in accordance with CSR reporting guidelines, whether reporting principles have been fulfilled and whether CSR minimum requirements have been met. Assessors check information given in reports in situ should a company have more than four full-time employees and certificates are checked on a random basis. In the second step an assessment is written following the verification which summarizes feedback, including advice for improvement as well as a comparison of the company with industry oriented CSR key indicators. Qualified feedback and key indicators allow the company to see its comparison with companies and to recognize the potential for improvement. The third and final step involves the council for certification, occupied by experts, deciding whether a certificate can be awarded on the basis of the CSR report and expert assessment at hand.

In general the first recertification takes place two years later and is then repeated in three year intervals. (Tourcert, 2011)

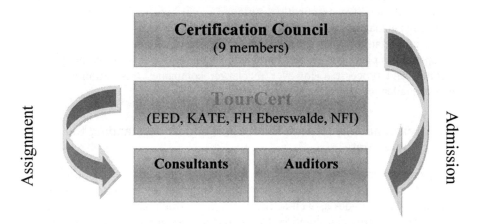

Fig. 4. The way to certification

Source: Tourcert, 2011

2 Effects on Travel Agents from Introducing the CSR System

2.1 Analytical Phase: Structured Self-Reflection

The CSR process of Tourcert is a special kind of company analysis: By getting an idea of the company's ecological and economic structure, questioning employees,checking products and service providers, who play a part in the value-adding chain, and finally analyzing client feedback allows tour operators to get a better insight into their own business. Especially self-reflection requires a lot of commitment and confidence – in particular from the managements point of view which has to be 100% behind the idea, the introduction and the amount of work which it demands.

Starting off surely poses the greatest hurdle: First of all, incorporating the Tourcert system demands that internal ways of collecting and compiling data efficiently are developed. The first implementation of the CSR process takes much longer than all re-certifications which follow later on. Data collection and product analysis is team work; involving every employee is absolutely necessary. This guarantees the flow of information, everyone's mental participation and also assisting in developing ideas and implementing improvements.

A lot of time needs to be invested should the system be wholeheartedly installed into the company and own work is scrutinized. Being a small company with eight employees the hamburgian tour operator a&e erlebnis:reisen has nevertheless managed to analyse 85% of its self-produced core business, regarding products and CO^2 emissions as well as evaluating 41 partner agencies and 189 accommodations using the CSR system. Collecting these data forces being systematic and can, if consequently employed, contribute to improve the internal knowledge management.

It is still relatively easy for a company the size of a&e erlebnis:reisen to carry out specific implementation and knowledge transfer due to short communication channels and processes. However for larger companies this certainly poses a greater challenge.

2.2   Sustainability Report: Evaluating and Understanding the Actual State

Analyzing collected data is a requisite for creating the standardized sustainability report. Results of different checks performed on accommodations, offers, partner agencies or tour-guides as well as precise figures which have been raised regarding company energy consumption, CO^2 emissions per guest and also the customer and employee satisfaction index provide detailed information about the actual state of the company. This provides an enlightening moment especially for tour operators who have actively taken part in sustainable tourism for many years. The CSR process has made it possible for the first time to measure and verify the performance of a tour operator in terms of sustainability and provides an answer to the question: How sustainable are we really?

Verbalizing the results of the sustainability report provides the opportunity for the company to consciously illustrate and describe its own actions in detail. Tables and diagrams allow a quick, ascertainable and clear illustration of information.

2.3   Improvement Program as an Instrument of Quality Management

An important requirement of the CSR process is creating an improvement program within the framework of the sustainability report. For this aspect, the participation of all employees is crucial. The analysis allows strengths and weaknesses of a company to be recognized, described in detail and tackled. This works best if all team members sit down together and develop ideas, where potential for optimization can be found and how this could be implemented. Be it single internal processes which can be simplified, opportunities to save energy and material or changing individual products – collectively taking part allows suggestions and impulses, as well as motivation to put thoughts of improvement immediately into practice. Whoever is enthusiastic about something works with more motivation. This in turn is the basis for creative ideas, for future development as well as a high productivity. This guarantees a company's quality and success.

With plans for improvement fixed in writing the company makes the aspired optimization transparent – internally for employees as well as externally for clients and Tourcert assessors. Implementing the self-imposed improvements has to take place in clearly specified periods. This is why the own yardstick is determined and can be understood, measured and checked.

2.4 Communication: Certification as Recognition and Publicity as a Regulator

External verification follows self-analysis and insight – this also motivates and spurs on to collect data in full and in a comprehensible way. Verification by independent assessors and certification by an external council for certification is public recognition that sustainability is embedded in the company. This assures clients that the company's ways of acting in the market are credible.

By publishing the report and all information which it contains as well as the self-imposed improvement program the tour operator openly shows his sphere of action. Clearly detailed information in a standardized structure is provided. The freely accessible sustainability report offers clients the opportunity of verification. Who doesnot implement what is promised can be warned by external regulative forces. In public relations, the CSR seal can be included as a supporting confidence-building measure for clients. Information contained in the report also helps to actively sensitize clients in specific areas of sustainable travel.

Communicating results and completed analyses, as well as the opportunity for the tour operator to clearly position himself, is the greatest gain of the CSR process. Serious reporting allows the company to show that sustainability requirements are not just "flash in the pan" marketing but rather a fundamental guiding principle and firmly established, lived business culture.

However, communication is not just important for the relationship with clients. Being open towards one another is a requisite for a successful completion as can clearly be seen when going through the CSR process. This concerted approach on the one hand refers to processes being done in the office by management and employees as well as among colleagues. On the other hand, communication and handling all participating business partners fairly, in Germany as well as worldwide, is of great importance. Replies made by local agencies and partners reflect the great interest shown in sustainability and the CSR reporting process. Talks about requirements and analysis results enable all service providers to be included and new suggestions arise for future cooperation, which continue through into product development.

2.5   Sustainability as a Process: Continually Optimizing and Developing

The continuous development of a company with regards to its sustainable qualities is one of the main aspects of Tourcert's CSR process. A long-term implementation is very important. Not just the way of travelling should be sustainable, but the certification also has sustainable effects on processes within a company. Re-certification at regular intervals is a logical part of a lasting sustainable optimizing process, which ensures that sustainability does not remain a static observation but rather triggers a dynamic process, which contributes to continuous improvement.

References

Forum anders reisen: criteria catalogue: http://forumandersreisen.de/mitglieder_kriterienkata-log.php 23.07.2011 (2004)

Hardtke, A. & Kleinfeld, A. (ed.): Gesellschaftliche Verantwortung von Unternehmen. Von der Idee der Corporate Social Responsibility zur erfolgreichen Umsetzung. Wiesbaden: 2010

International Organisation für Standardization (ISO): ISO 26000 – Social responsibility, http://www.iso.org/iso/iso_catalogue/management_and_leadership_standards/social_re-sponsibility/sr_iso26000_overview.htm 22.07.2011 (2011)

Tourcert: Corporate Social Responsibility in Tourism, http://www.tourcert.org/index.php?id=csr-prozess&L=0 21.07.2011 (2011)

Challenges Awaiting the Aviation Industry – Preparation for the Integration in the Emission Trading Scheme

Dietrich Brockhagen

1 Introduction

As of 2012, the aviation sector will be integrated into the EU Emission Trading Scheme (EU ETS). However, some airlines, aircraft manufacturers and governments are still trying to deter the European community from implementing the directive. Heated conflict seems to be programmed, given the EU –Commission's firm stance: "The inclusion of aviation in the ETS is not a proposal, it is now European law. [...] So we are not thinking at all about the possibility of changing our legislation." said European Commission President Jose Manuel Barroso in June 2011[1].

In the past, global aviation benefited from several tax exemptions regarding VAT and kerosene. Inclusion in the EU ETS will be the first significant international effort to internalize external climate costs following the polluter-pays-principle. Whereas airlines have a natural economic incentive to reduce kerosene consumption and thus CO_2 in order to save costs, the inclusion in the EU ETS puts additional pressure on the industry to reduce the CO_2 emissions from flights.

This article investigates some available options for the aviation industry which reduce CO_2 emissions from flights and thus lower costs for compliance with the EU ETS. Section 4 will take a closer look at cost reducing synergies with voluntary CO_2 offset schemes.

2 Main Factors Causing CO_2 Emissions in Flight Operation

At the ITB 2011, the German climate business atmosfair launched an index examining the climate efficiency of the 130 biggest airlines in terms of CO_2 per revenue kilometre. Airlines are attributed from 1 to 100 efficiency points and assigned to

---

[1] Reuters (2011).

7 efficiency classes from A to G, similar to the EU energy efficiency label. For the allocation of efficiency points, only CO_2 and NO_x are considered (no noise, no sustainability policy etc.). The atmosfair Airline Index (AAI) is based on the CO_2 calculation methodology of ICAO, but refines the input and output parameters. It uses exclusively high-ranking sources from international organizations, such as ICAO or IATA or dedicated industry service providers such as OAG. According to Paul Peeters, aircraft engineer and professor for sustainable tourism and transport at NHTV Breda University, "the AAI calculation method is precise and sets the standard for the environmental evaluation of aircrafts and airlines."[2]

The index reveals the main factors causing specific CO_2 emissions of flights. Figure 1 shows the result of the analysis of more than 22 million flights on 13.000 city pairs. The figure is based on the following approach, regarding the seven most important factors influencing specific CO_2 emissions: If one changes one factor by an industry typical standard deviation and keeps all other six factors constant (ceteris paribus), the numbers indicated in the figure show, how much the specific CO_2 emissions change as a result for changing this factor.

By this account, obviously, the passenger load factor (48%) is the most important lever for reducing the climate impact of airlines, followed by aircraft type (31%) and seating capacity (8%).[3]

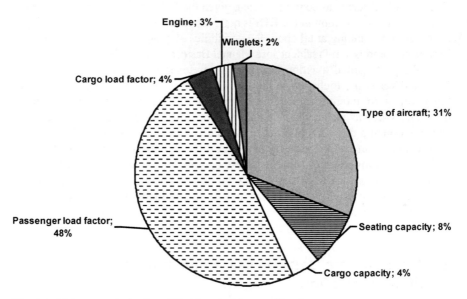

Fig. 1. Efficiency optimization: What has the biggest effect?

Source: Brockhagen et al. (2011): 88

---

[2]   Brockhagen et al. (2011): 5ff.
[3]   Ibid., 28ff.

3 Optimizing the Climate Efficiency of Flights: Detailed City Pair Analyses for Airlines and Companies

The AAI was developed in order to ensure transparency of the climate impact of aviation while supporting the airline industry in reducing its climate footprint. For this purpose, the index serves as an assessment tool which assists airlines in optimizing their climate efficiency and identifying emissions reduction potentials. atmosfair delivers the AAI in detail for single routes. Two products are available for airlines:

Airlines and efficiency rank	Climate efficiency [points]	Average ticket price [EUR]	Change assessment [eff./price]	Net load factor (pax & cargo) [%]	Seating capacity [seats]	Kerosene consumption [kg]
Potential optimum Boeing 777-200LR	100	-	-	100,0%	440	42.498
1. Continental Airlines	60,2	778	+/+			
Boeing 757				54,8%	175	20.784
Boeing 777				39,7%	283	35.900
CodeSharing:						
Lufthansa		785	+/o			
United Airlines		769	+/+			
British Midland		785	+/o			
Iberia		954	+/-			
2. Delta Airways	54,8	1001	+/-			
Boeing 767-400				53,3%	281	31.839
Boeing 767				39,4%	214	28.770
CodeSharing:						
KLM		1010	+/-			
3. British Airways	53,7	785	current			
Boeing 767				54,6%	216	31.150
Boeing 747-400				48,1%	291	59.570
Boeing 777				47,8%	267	39.140
CodeSharing:						
Iberia		954	o/-			
4. American Airlines	44,9	788	-/-			
Boeing 777				47,8%	247	40.486
CodeSharing:						
Jet Blue		769	-/+			
5. Virgin Atlantic	43,3	785	-/o			
Boeing 747-400				64,2%	451	56.327
Airbus A340-600				51,4%	308	43.382
Airbus A340-300				43,2%	240	36.800

Change assessment		
	\\\\\\\\\	Better climate efficiency and not more expensive than current airline
+ win		
- loose	▓▓▓▓▓	Current airline
o unchanged	/////////	Worse climate efficiency and not cheaper than current airline

Fig. 2. atmosfair airline index report sample: Assessing hypothetical changes from the current airline (in this case: British Airways) to another airline with regards to climate efficiency and ticket price.

Source: atmosfair gGmbH

- **Competitive reporting** offering detailed data on various factors (load factors, fuel consumption etc.), comparing all airlines operating on a specific city pair.

- **Tailor-made reporting** delivering a sensitivity analysis for the dependency of climate efficiency points from hypothetic factors set for a given aircraft on a specific city pair.

Additionally, atmosfair offers the AAI for individual routes to corporate clients. Many companies have already implemented awareness-raising travel management systems which encourage employees to choose more climate-friendly means of transport whenever reasonable. By identifying the most climate-efficient airlines for the most important routes, atmosfair supports cost-conscious and environmentally aware companies in reducing their business travel CO_2 emissions.

Companies that launch RFPs for airline services on specific routes may use the AAI to add climate efficiency as a criterion. atmosfair delivers an assessment which analysis how by switching airlines they can leverage a win-win potential with regard to both climate efficiency and ticket price. Figure 2 shows an AAI working sample for the city pair London-New York. It evaluates the hypothetical change from the currently used airline to other airlines.

4   Investment in Company-Owned Offset Projects and Voluntary Offsetting by Passengers: Multiple Benefits for Airlines

In order to comply with the EU ETS, airlines will have to surrender emission allowances for their CO_2 emissions of all flights to and from the EU. The available allowances (EUAAs) will be limited to 97 percent (2012) and 95 percent (2013–2020) of the average sector emissions of 2004–2006. The growth in transport volumes of the aviation sector and the associated growth in emissions require additional allowances from other sources.[4]

Measures to avoid and reduce emissions from flight operation will most likely not be sufficient to compensate for the additional allowances needed. For that reason, most airlines will have to procure emission certificates, either from other EU-sectors (EUAs) or from offset projects in developing countries (CDM-Projects, CERs). Given the fact that EUA prices are predicted to grow strongly over the course of the next years, it is worth taking a closer look at emission certificates which originated from offset projects.

Apart from buying CERs from the secondary CDM-market, airlines have the option to invest directly in offset projects in order to generate their own CERs.

---

[4]   European Parliament (2008): Article 1ff.

atmosfair currently develops several efficient cook stove projects in least developed countries which are ready to take up investments by airlines. A similar project was already implemented successfully for German cargo carrier DHL by atmosfair. Since the airline becomes the owner of the project, it also bears the risk of project failure. On the other hand, such an investment reaps multiple benefits, such as:

- **Savings:** An airline-owned cook stove project may lead to substantial savings and price stability compared to purchased allowances. Such projects can yield CERs at calculatory prices of below 10 EUR constantly, whereas EUA and secondary CER prices are forecasted to increase to >20 EUR by 2015[5].

- **Safety:** The EU will restrict the eligibility of CERs generated in projects which have low benefits regarding sustainable development or are registered after 2013[6]. Preferable are thus CDM Standard projects located in Least Developed Countries or registered before 2013 and thus safely eligible for the EU-ETS until 2020.

- **Flexibility:** An airline-owned offset project leads to valuable benefits in the sourcing process of CERs, such as secure access and scalability.

- **Marketing effects:** In comparison to purchased allowances, the investment in airline-owned projects allows the airline for an authentic B2C communication and is a credible pillar within its CSR strategies.

In addition to the benefits mentioned above, airlines have the opportunity to harness voluntary offsetting of passengers for lowering compliance costs. The setup of the CDM project and some overhead costs can be financed by the passengers, thus minimising the risk for the airline. All certificates gained with their offset money will be used for CERs that are retired for the benefit of the atmosphere, thus carrying out the offset of the passengers" emissions. Once set up and running, the airline can expand the CDM project with its own money in order to generate its own CERs for compliance with the EU – ETS. This approach lowers the CER generation costs of the airline.

Figure 3 shows how airlines can start easily and at no cost and no risk: They integrate the ready to use atmosfair offset box in its online sales systems. This symbolizes the airline's commitment to save the climate while customers have the chance to voluntarily support the work of the airline.

An airline owned offset project in developing countries requires an investment over a few years. This results in a project generating CERs for at least 10 years. Combining the generation of CERs for compliance with the integration of voluntary offsetting in the airline's booking channel means a triple-win situation:

---

[5] Barclays Capital (2011).
[6] European Parliament (2009): Article 11a.

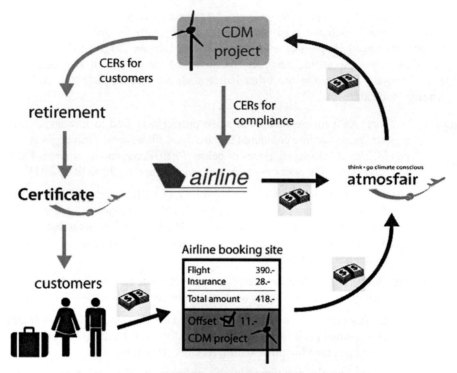

Fig. 3. Combining voluntary offsetting and compliance with the EU-ETS.
Source: atmosfair gGmbH

- **Benefit to the airline:** A successful implementation of a cook stove project ensures low-priced CERs, ETS-eligibility, scalability in the sourcing process, the absence of price fluctuations and a credible B2C communication in the CSR area.

- **Benefit to environment and people:** CDM Gold Standard projects are trustworthy to the public. They guarantee that CO_2 emissions are reduced to the indicated extend. Furthermore, the projects meet high standards regarding job creation, health, income, gender mainstreaming and promotion of advanced technologies.

- **Benefit to the passengers:** The integration of the atmosfair offset box into the booking process constitutes a quality element and gives passengers the opportunity to contribute to climate protection voluntarily.

5 Conclusion

Integration into the EU ETS is not a ground known to airlines. Consequently, the latter will seek the most suitable solutions regarding cost reduction and planning security. The atmosfair Airline Index is a suitable tool assisting an airline to assess its main factors causing CO_2 emissions and its competitive position in the CO_2 arena.

With stable forecasts for growing CO_2 certifcate prices over the coming 10 years of the EU ETS, airlines should seek strategies that combine planning security and cost reductions. The investment in own CDM projects may lead to low-priced emission certificates which are safely eligible in the EU ETS. Airlines have a special advantage here, akin only to this industry: The combination with voluntary offsetting schemes minimizes the initial costs and risks for the airline.

References

Barclays Capital (2011): Carbon Market Outlook, March 2011.

Brockhagen, D. et al. (2011): Airline Index 2011 – Documentation of the methodology, Berlin.

European Parliament (2008): EU Directive 2008/101/EC of November 19th 2008 amending Directive 2003/87/EC

European Parliament (2009): EU Directive 2009/29/EC of April 23rd 2009 amending Directive 2003/87/EC

Reuters (2011): EU's Barroso stands firm on ETS for airlines, June 8th 2011, www.reuters. com

Corporate Social Reporting and Practices of International Hotel Groups

Xavier Font and Andreas Walmsley

1 Taking Responsibility for Sustainability in the Hospitality Industry

There is increasing pressure for corporations to disclose their sustainability management practices. Often this takes place through corporate social and environmental reports, and companies leading on good CSR practice now allow access to an audit of the information contained within these reports. A recent KPMG study (KPMG International, 2008) has shown that CSR reporting is becoming the norm in large corporations. Whereas in 2005 only 50% of companies included CSR in their reporting, the figure for KMPG's 2008 report stands at almost 80%.

It is not as though CSR is without its critics. There is much debate on whether the corporate social reports of companies are a form of posturing, or actually reflect improved environmental and social management (Adams, 2004; Al-Tuwaijri, Christensen and Hughes, 2004; Beattie, McInnes and Fearnley, 2004; Clarkson *et al.*, 2008). At present there is no agreement in the literature, some studies have found that increased disclosure does reflect improved performance, while others have found that it only reflects improved management systems- ideally the management systems would be a means to an end (performance) but this is not always the case, and there's too much evidence of CSR reports being a form of corporate posturing (Laufer, 2003). Evidently, CSR reporting should be about more than solely promoting a company's positive initiatives and actions.

Hess (2008) provides a further critique of CSR arguing that too often it is implemented by companies in an attempt to discourage policy makers from introducing further regulation. According to Hess (2008), not only must the corporation disclose material information and engage with its stakeholders, but it should also become a more moral organisation. Whether the firm becomes more moral in its values or not, the implementation of CSR systems, the reporting of CSR activities and the subsequent independent auditing of the CSR reports should lead to improved CSR performance. The latter point about independent audits is critical in this endeavour of achieving actual improved CSR performance as opposed to pure posturing.

It has to be acknowledged however that what to include in a typical CSR report is open to debate. In fact, while we use the term CSR here, a profusion of terms surrounding the broad notion of responsibility in business exists, e.g. business ethics, corporate social performance and corporate governance. This hampers the comparability of firms" CSR performance as different firms may well focus on different aspects of CSR (Paul, 2008). Nonetheless, guidelines such as those provided by the Global Reporting Initiative are making the task as to what to include in a CSR report somewhat easier, as well as enabling comparability across businesses.

The literature on CSR reporting on hotels is weak. Much of the literature seems to take a normative stance, attempting to make the business case for hotels to engage in environmental management specifically (Álvarez Gil, Burgos Jimenez and Céspedes Lorente, 2001; Ayuso, 2006; Ayuso, 2007; Bohdanowicz, 2005; Bohdanowicz, 2006; Bohdanowicz, 2007b; Bonilla-Priego and Avilés-Palacios, 2008; Calveras, 2003; Carmona-Moreno, Céspedes-Lorente and De Burgos-Jimenez, 2004; Chan and Wong, 2006; Claver-Cortés et al., 2007; IHEI, 2005). Very little is said about the CSR aspects of human resource management (ILO, 2001) which is interesting in that early definitions of CSR focused heavily on the social as opposed to the environmental aspects of a company's responsibility (Morimoto, Ash and Hope, 2005). It is telling, and yet not too surprising given the sensitive nature of the topic that to date, the only hotel group that has allowed any academic disclosure of its CSR reporting is Hilton Europe (Bohdanowicz, 2007a; Bohdanowicz, Zientara and Novotna, 2011).

Despite what was said above about businesses often only reporting the positive aspects of their CSR behaviour, not all hotels report on the positive initiatives and actions that they engage in. This was evidenced by Merwe and Wöcke (2007) in a sample of South African hotels. Here it was found that28% of businesses in their sample had not considered including CSR activities in marketing material. It is not as though the call for increased CSR reporting is necessarily detrimental for firms, far from it. Many examples of positive behavior exist at the individual hotel level. The point is to turn the exceptional into the common. CSR reporting and auditing are key in achieving this.

2   Methodology

This study was commissioned by the organisation "International Consumer Research and Testing" on behalf of its members (henceforth Consumer Association). The purpose was to inform consumers of the impacts of tourism and influence their selection of hotel groups, as well as to show different corporate policies and how they are translated to practice. It aims to use consumer pressure to change industry behaviour and to provide a single comparable analysis of practices. Despite the emphasis on consumer pressure, it is hoped that in terms of providing information on hotels" CSR performance and reporting this will benefit the hotels themselves. Indeed, because of the different reporting mechanisms the comparability

ACCOR INTERNATIONAL BARCELÓ HOTELS & RESORTS CARLSON GROUP HILTON IBEROSTAR INTERCONTINENTAL HOTELS GROUP (IHG) MARRIOTT INTERNATIONAL RIU HOTELS SOL MELIÁ STARWOOD HOTELS&RESORTS	

Fig. 1. Hotel groups surveyed

of CSR reports has in the past been hampered. This study addresses this issue by providing a common method of assessment across ten leisure hotel groups that are of importance in particular to the European market.

The emphasis on this study was to research the corporate level policies that top leisure hotel groups have developed, and to then test how these are applied in a sample of hotels. Figure 2 shows the range of issues included in this research. These were developed in association with the Consumer Association.

CORPORATE POLICIES	• Endorsement of International key conventions • Resources for CSR • Staff training program on sustainable issues • CSR management systems • Independent certification of sustainability practices
LABOUR	• Policy on working conditions
SOCIO ECONOMIC	• Ethical, green and local purchasing policy • Policy on social impacts at the destination
ENVIRONMENTAL	• Environmental policy • Carbon footprint monitoring
CUSTOMER ENGAGEMENT	• Accessibility for wheelchair customers • Dietary needs • Customer sustainability education
TRANSPARENCY	• Cooperation with the survey • CSR reports

Fig. 2. Scope of CSR survey

The primary research for this study took six months and was conducted in four steps:

Step one involved policy analysis. We analysed information on company websites as well as internal documentation provided by the corporations, signing non disclosure of confidential information when required. We also reviewed publicly available information on controversies surrounding the corporations (legal cases where companies have contravened their CSR policies).

The second step assessed head quarter engagement. We conducted a survey of CSR practices in the different hotel groups, over a three month period. One hotel group submitted a very poor and incomplete response (Iberostar), while RIU hotels declined to participate altogether.

Step three comprised field visits to assess how individual hotels within these groups implement the corporate level policies in two or three locations, in Southern Europe, Mexico and Thailand. Three experienced assessors conducted staff interviews and observed CSR practices lasting between 6 and 24 hours in each hotel. Hotels were notified about the visit via their headquarters but at short notice.

The fourth and final step involved providing feedback to the hotel chains with the aim of improving the study's reliability. We sent the results comparing the corporate policies with individual hotel practices to each hotel group, to provide the opportunity to clarify anomalies. We conducted a detailed scoring assessment on policy and on practice, developing a grading system for policies, and using the field validation to then invalidate, partly validate or fully validate the results. The hotel groups did not receive the numerical grading information, just the textual explanation of which evidence was found, for the aim was not to argue on the scoring mechanism, but on the veracity of its underpinning data.

3   Results

The first section of the results reviews corporate policies to manage the sustainability process. The research showed that many hotel groups endorse international conventions (1. UNWTO Global code of ethics, 2. ILO Declaration on Fundamental Principles and Rights at Work, 3. UN Global Compact, 4. Universal Declaration of Human Rights and 5. OECD Guidelines for Multinational Enterprises). However, in visiting the hotels, it was evident that this did not always translate into action. The human resources function dedicated to CSR/sustainability in each hotel varied considerably. While nearly all hotel groups had a CSR nominee, this was in many cases a tokenistic addition to someone else's current role, and no time was made available for this additional role. CSR responsibilities were most often aligned with the role of chief engineer. Several companies had a staff training program that covered environmental issues for operational staff (not necessarily for management), and few of them looked into socio-economic/ethical issues. Where the latter did exist, this revolved around ECPAT training on sexual exploitation of minors. The management systems data for CSR purposes was variable. Some hotel groups didn't collect any data for that purpose, while the most common data to be found were on energy and water consumption. The company's management systems did not include social/ethical issues following relevant guidelines such as OHSAS 18001, SA8000, AA1000 or tourism specific standards. A range of labels such as ISO 14001, EMAS, Green Globe 21, Green Tourism Business Scheme and others were used by some of the hotels in different groups, but the only group to have a company wide CSR management system was Sol Meliá, with Biosphere Hotels.

The second section of the survey focused on labour issues. There was evidence that the ILO core conventions were covered by the working conditions policy (Forced Labour, Freedom of Association and Protection of the Right to Organize, Right to Organize and Collective Bargaining, Equal Remuneration, Abolition of Forced Labour, Discrimination in Employment and Occupation, Minimum Age Convention, and Elimination of the Worst Forms of Child Labour were all included). Wages were covered by the working conditions policy for minimum (legal) wages, but not in regards to living wages. Working conditions were covered by the working conditions policy (in Health & Safety, Maximum hours of work/overtime, Disciplinary practices, and Gender/family friendly working policies such as pregnancy, child care). Discrimination was covered by the working conditions policy in the legal sense of the definition, regarding sex, disability, age and race discrimination or policies to prevent sexual harassment. There were however no positive employment policies for example on hiring local people. While the paper trail of Human Resource Management was well covered, it was obvious that it is not integrated within CSR. Local legal compliance was given as the standard answer (in many places negotiated by labour unions) with little effort to go beyond these arrangements.

Third, the inclusion of socio-economic issues outside the physical boundaries of the hotel was rather disappointing. The purchasing policies for these companies rarely covered selection of locally produced, Fair-trade-labelled or eco-labelled goods, and there was little evidence of understanding or attempting to manage the impacts caused by the hotels through their supply chains. Marriott and Intercontinental monitor any loss of natural resources, but there is little evidence of monitoring access by local communities to resources (i.e. beaches, water, wood, fuel, fishing rights). The contribution to socio-economic issues affecting the destination as a whole tends to happen on a hotel by hotel basis. Group level interventions, other than in support of philanthropic or community projects, are rare. Philanthropy still dominates how hotel groups tackle CSR issues at a local level.

The fourth aspect was environmental management, central to the CSR policies of all hotel groups. Water, energy and waste management were all widely implemented. To a certain extent this shows current inefficiencies in the management of buildings, coupled with a culture of luxury consumption and waste. Solid waste management is taking place primarily through local government pressure to reduce landfill (but not gone upstream, the requirements have not been passed on to suppliers to reduce packaging). Other environmental aspects are less widespread. Strong biodiversity policies are rare (Accor, Carlson Europe, and Inter Continental), and most groups fail at the implementation phase. Carbon footprint monitoring takes place as an extension of the energy management policy, and only in the larger groups, while carbon calculations are limited to in house emissions not on supply chains.

The fifth aspect of our research, called customer engagement, was predicated on two types of variables. Issues of inclusivity were considered important, such as the provision of facilities for people with disabilities or with special dietary needs. In practice these were poorly covered by most hotels and the research was limited to wheelchair access and celiac intolerances for being the most common inclusion aspects, and yet even here the results were disappointingly low. This was followed by communicating sustainability practices to customers and engaging them in supporting the company's efforts (such as recycling). Accor, Carlson Europe and Asia-Pacific, Inter Continental and Marriott go further than most, but besides tokenistic "towel agreement" efforts, and some generic philanthropy communications, the general approach was one of not disturbing the customer.

The final aspect was one of transparency, understood in two ways. First, as co-operation with the survey (all chains participated except RIU, and Hilton did not allow a visit to one of its hotels (for which they were severely penalised in their overall score). Secondly, as the publication of corporate social responsibility reports (where most groups have limited public reports, Accor has a comprehensive report, and Carlson Europe (but not Americas or Asia-Pacific), Inter Continental, Marriott and Sol Meliá have Global Reporting Initiative checked reports.

4   Discussion

The ten hotel groups were compared in their corporate social responsibility policies, validated through field research. Our research has found that these hotel groups are at clearly different stages of development and implementation of corporate policies. We found evidence that individual hotels tend to go beyond corporate policy on some aspects, but overall there was no consistency, little group management, and corporate policies are usually behind, not ahead of hotel behaviour. The results of this benchmarking are presented in figure 3 followed by some reflections.

Environmental management is central to how CSR is perceived by hotel groups, as a form of reducing costs through the many eco-savings available. The savings that can be made through water and energy management are substantial, both through small investments with reasonable payback periods, but also importantly through staff training (Bohdanowicz, Zientara and Novotna, 2011). However in practice, quality and health and safety prevails when there's a perception of potential conflict (Baddeley and Font, 2011) and there is little willingness to change practices that might be seen by customers as impacting on their luxury indulgence

		CORPORATE	LABOUR ISSUES	SOCIO ECONOMIC	ENVIRONMENTAL	CUSTOMER	TRANSPARENCY	Total (0–100)
1°	ACCOR	B	B	B	A	B	A	79
2°	SOL MELIÁ	C	B	B	B	C	A	66
3°	MARRIOTT	C	B	B	B	C	A	66
4°	CARLSON	C	B	B	B	C	B	65
5°	IHG	C	A	B	C	D	A	64
6°	STARWOOD	C	B	C	A	D	B	64
7°	BARCELÓ	C	C	C	C	C	A	52
8°	HILTON	C	D	D	C	D	B	41
9°	IBEROSTAR	E	E	D	D	D	B	31
10°	RIU	E	E	E	E	E	E	6

Fig. 3. Benchmarked results

Source: Konsument, March 2011

experience, as seen by the limited willingness of communicating sustainability actions, let alone asking customers to behave in a more sustainably appropriate way.

The content of the policies analysed was mostly inward looking, focusing on operational efficiencies with little acceptance of impacts on the destination. We have found no hotel group that use the eco-savings from water and energy efficiency to create CSR budgets for issues that will inevitably require increasing costs, or at least short term investment. Equally we have seen little evidence of linking human resource management policies and practices to CSR which seem to be limited to encouraging environmental effectiveness (Bohdanowicz, Simanic and Martinac, 2005; Bohdanowicz, Zientara and Novotna, 2011). The noise made about philanthropy pales into significance when it is put in the context of the positive economic and social impact that hotel groups could make to local societies if they changed marginally their employment and purchasing policies to encourage more sustainably made, locally sourced goods and flexible working conditions for local people with difficulties to find employment. Sustainable supply chain management must become part of how large hotel groups see themselves to contribute to sustainable development, despite the obvious challenges in securing the volumes, quality and servicing required (Mamic, 2004; Tan, 2001; The National Environmental Education & Training Foundation, 2001).

The larger hotel groups generally have better defined CSR policies and documentation. The two exceptions would be Sol Meliá and Hilton- the first is relatively small with less than 300 hotels but has a well designed internal tracking system to measure performance against a comprehensive CSR policy. The latter is one of the largest hotel groups worldwide, yet has focused on eco-savings-linked environmental management, with limited integration of CSR into human resources policies or other aspects of the business. International hotel groups are far more strategic in their environmental management behaviour than the smaller ones, with well defined building management systems and group wide benchmarking methods- domestic hotels of the same size rarely will have such systems despite the potential for substantial savings (El Dief and Font, 2011). Beyond that, there is evidence of lack of strategic thinking in the data collected in two accounts, the above mentioned potential for internal efficiencies, and the increasingly important requirements from distribution channels and the tourism markets (Bonilla-Priego, Najera and Font, 2011). In recent years we have heard much about the CSR policies of tour operator groups, TUI in particular. Yet it is important to highlight that RIU hotels is 50% owned by TUI, and their poor performance coupled with their unwillingness to participate undermines the TUI group policy.

5   Conclusions

In 2009 Pizam asked whether "green" or "sustainable" hotels were a fad, ploy or fact of life. Answering his own question he states: "in the middle and long-term future of the hotel industry, it is my personal belief that … like most other businesses [hotels] will have no other choice but to become „real green'" (Pizam, 2009) It is true that on the basis of this study hotel chains are engaging in the environmental side of CSR, although more could of course be done. What appears less to be the case is a serious engagement with the non-environmental elements such as employee relations and the socio-economic impacts within the destination.

At a time of increasing public awareness and hence scrutiny of corporate claims, be they financial or CSR related, it is in the corporation's best interest to be open and honest about what it does. As the KPMG (2008) study indicates, a growing number of global businesses are engaging in corporate social reporting and are willing to be scrutinised on the basis of these reports. While deciding what to include in such reports is admittedly no easy task, guidelines such as the Global Reporting Initiative exist to assist the CSR manager.

Perhaps the hotel industry is unique in its high proportion of franchised businesses. Nevertheless, it would be unreasonable to put the industry's CSR performance down to a lack of corporate control over its assets. In as much as the hotel chain provides policy at a corporate level on financial performance and marketing, for example, so it should provide these guidelines in terms of CSR. More importantly, it should then assist individual hotels in ensuring these guidelines are adhered to. Hotels are part of destinations, their responsibilities extend beyond their own walls.

Acknowledgements. We are grateful for International Consumer Research and Testing commissioning this study. The results have been published in March to July 2011 in nine consumer association magazines:

- Austria: Verein Fur Konsumenteninformation – http://www.konsument.at
- Belgium: Association Belge des consommateurs Test-Achats – http://www.test-achats.be/
- Denmark: Taenk/Forbrugerraadet (Danish Consumer Council) – http://www.taenk.dk/
- Finland: Kuluttajavirasto – http://www.kuluttajavirasto.fi/
- Italy: Euroconsumers Servizi SRL – http://www.altroconsumo.it/
- Portugal: DECO-Proteste Editores LDA – http://www.deco.proteste.pt/
- Spain: OCU-Organización de Consumidores y Usuarios Ediciones SA – http://www.ocu.org/
- Sweden: Rad & Ron – http://www.radron.se/
- Switzerland: Fédération romande des Consommateurs – http://www.frc.ch/

References

Adams, C. (2004) The ethical, social and environmental reporting performance portrayal gap. *Accounting Auditing and Accountability Journal* 17 (5), 731–757.

Al-Tuwaijri, S. A., Christensen, T. E. and Hughes, K. (2004) The relations among environmental disclosure, environmental performance, and economic performance: a simultaneous equations approach. *Accounting, Organizations and Society* 29 (5–6), 447–471.

Álvarez Gil, M. J., Burgos Jimenez, J. and Céspedes Lorente, J. J. (2001) An analysis of environmental management, organizational context and performance of Spanish hotels. *Omega* 29 (6), 457–471.

Ayuso, S. (2006) Adoption of Voluntary Environmental Tools for Sustainable Tourism: Analysing the Experience of Spanish Hotels. *Corporate Social Responsibility and Environmental Management* 13 (4), 207–220.

Ayuso, S. (2007) Comparing Voluntary Policy Instruments for Sustainable Tourism: The Experience of the Spanish Hotel Sector. *Journal of Sustainable Tourism* 15144–159.

Baddeley, J. and Font, X. (2011) Barriers to Tour Operator Sustainable Supply Chain Management. *Tourism and Recreation Research* in publication.

Beattie, V., McInnes, B. and Fearnley, S. (2004) A methodology for analysing and evaluating narratives in annual reports: a comprehensive descriptive profile and metrics for disclosure quality attributes (pp. 205–236): Elsevier.

Bohdanowicz, P. (2005) European Hoteliers" Environmental Attitudes: Greening the Business. *Cornell Hotel and Restaurant Administration Quarterly* 46 (2), 188–204.

Bohdanowicz, P. (2006) Environmental awareness and initiatives in the Swedish and Polish hotel industries-survey results. *International Journal of Hospitality Management* 25 (4), 662–682.

Bohdanowicz, P. (2007a) A case study of Hilton Environmental Reporting as a tool of corporate social responsibility. *Tourism Review* 11 (2), 115–131.

Bohdanowicz, P. (2007b) Theory and Practice of Environmental Management and Monitoring in Hotel Chains. *In* S. Gossling, C. M. Hall, & D. B. Weaver (Eds.), *Expert meeting on sustainable tourism; Sustainable tourism futures : perspectives on systems, restructuring and innovations* (pp. 102–130). Helsingborg, Sweden: New York.

Bohdanowicz, P., Simanic, B. and Martinac, I. (2005) Environmental training and measures at Scandic Hotels, Sweden. *Tourism Review International* 9 (1), 7–19.

Bohdanowicz, P., Zientara, P. and Novotna, E. (2011) International hotel chains and environmental protection: an analysis of Hilton's we care! programme (Europe, 2006–2008). *Journal of Sustainable Tourism* 19 (forthcoming).

Bonilla-Priego, M. J. and Avilés-Palacios, C. (2008) Analysis of Environmental Statements Issued by Spanish EMAS-Certified Hotels. *Cornell Quarterly* 49 (4), 381–394.

Bonilla-Priego, M. J., Najera, J. J. and Font, X. (2011) Environmental management decision-making in certified hotels *Journal of Sustainable Tourism* 19 (3), 361–382.

Calveras, A. (2003) Incentives of international and local hotel chains to invest in environmental quality. *Tourism Economics* 9 (3), 297–306.

Carmona-Moreno, E., Céspedes-Lorente, J. and De Burgos-Jimenez, J. (2004) Environmental Strategies in Spanish Hotels: Contextual Factors and Performance. *Service Industries Journal.* 24 (3), 101–130.

Chan, E. and Wong, S. (2006) Motivations for ISO14001 in the hotel industry. *Tourism Management* 27481–492.

Clarkson, P. M., Li, Y., Richardson, G. D. and Vasvari, F. P. (2008) Revisiting the relation between environmental performance and environmental disclosure: An empirical analysis. *Accounting, Organizations and Society* 33 (4–5), 303–327.

Claver-Cortés, E., Molina-Azorín, J. F., Pereira-Moliner, J. and López-Gamero, M. D. (2007) Environmental strategies and their impact on hotel performance. *Journal of Sustainable Tourism* 15 (6), 663–679.

El Dief, M. and Font, X. (2011) Determinants of environmental management in the Red Sea Hotels: personal and organizational values and contextual variables. *Journal of Hospitality & Tourism Research* in publication.

IHEI (2005) Sowing the seeds of change. London: International Hotels Environment Initiative.

ILO (2001) Human resource development, employment and globalization in the hotel, catering and tourism sector. Geneva: International Labour Organization.

KPMG International (2008) KPMG International Survey of Corporate Responsibility Reporting 2008 (pp. 118): KPMG.

Laufer, W. S. (2003) Social Accountability and Corporate Greenwashing. *Journal of Business Ethics* 43 (3), 253–261.

Mamic, I. (2004) *Implementing Codes of Conduct: How Businesses Manage Social Performance in Global Supply Chains.* Sheffield: Greenleaf Publishing (UK).

Merwe, M. v. d. and Wöcke, A. (2007) An investigation into responsible tourism practices in the South African hotel industry. *South African Journal of Business Management* 38 (2), 1–15.

Morimoto, R., Ash, J. and Hope, C. (2005) Corporate Social Responsibility Audit: From Theory to Practice. *Journal of Business Ethics* 62 (4), 315–325.

Paul, K. (2008) Corporate Sustainability, Citizenship and Social Responsibility Reporting. *Journal of Corporate Citizenship* Winter 2008 (32), 63–78.

Pizam, A. (2009) Editorial: Green hotels: A fad, ploy or fact of life? *International Journal of Hospitality Management* 28 (1), 1.

Tan, K. C. (2001) A framework of supply chain management literature. *European Journal of Purchasing and Supply Management* 7 (1), 39–48.

The National Environmental Education & Training Foundation (2001) Going green, upstream: the promise of supply chain environmental management (pp. 55). Washingdon D.C.: The National Environmental Education & Training Foundation.

Eco-mobility

Status Quo and Future Prospects of Sustainable Mobility

Roland Conrady

1 Introduction

Climate change, CO^2 emissions, depletion of fossil fuels[1] and energy costs have been intensely discussed worldwide for many years. These topics can be called megatrends without having to exaggerate.

The transportation sector is one of the largest consumers of fossil fuels also making it one of the biggest producers of CO^2. Therefore political and private stakeholders are becoming increasingly aware of the transportation sector. Generally it can be said, that the entire transportation system is a phase of transition.

This article provides a general picture of progress being made by different means of transport in making transport more sustainable and shows probably trends.[2]

2 Resource Consumption and Climate Impact of the Transportation Sector

Transportation and fossil fuels are up to today strongly linked together: More than 60% of the 84 million barrels of oil which are consumed daily are used to power LDVs, HGVs, airplanes, ships and other means of transport.[3] In the future a further increase in consumption of fossil fuels is predicted (see passenger transport volumes in 17 Western European countries in Table 1, excluding ships). On a global basis the increase of transportation services in fast growing markets of Asia have a considerable impact.

---

[1] The statistical lifetime (which assumes constant consumption and constant output) for crude oil is 41 years, for natural gas 61 years, cf. Dena (2010), p. 14.

[2] Particular attention is drawn to the fact that status quo observations and forecasts are being made in the context of rapid change.

[3] Cf. World Economic Forum (2011), p. 8.

237

Table 1. Passenger transport volume in 17 Western European countries from 2008 to 2025. Source: Progtrans (2011).

Carrier	Total Traffic 2008 (in billion passenger kilometers)	Total Traffic 2025 (in billion passenger kilometers)	CAGR 2008-2025	Percentage 2008	Percentage 2025
LDV	4,284	4,552	+ 0.4 %	74.1 %	71.1 %
Bus	442	492	+ 0.6 %	7.7 %	7.8 %
Long-distance rail traffic	376	466	+ 1.3 %	6.5 %	7.3 %
Short-distance rail traffic	68	87	+ 1.5 %	1.2 %	1.4 %
Airplane	604	751	+ 1.3 %	10.5 %	11.8 %
Total	5,774	6,348		100.0 %	100.0 %

It is established that for a "continue as before"-scenario the transportation sector will consume 40% more energy than today by 2030 (see Table 2). [4] The transportation sector would thereby be responsible for 97% of the worldwide increase in oil consumption.[5]

Table 2. Worldwide energy consumption of the transportation sector 2010 – 2030 in a "continue as before"-scenario. Source: World Economic Forum (2011), p. 61.

Carrier	Energy consumption (in Mtoe/year 2010)	Energy consumption (in Mtoe/year 2030)	CAGR 2010–2030
LDV	1,144	1,667	1.9 %
Truck	379	464	1.0 %
Bus	81	66	–1.0 %
Aviation	215	411	3.3 %
Marine	227	272	0.9 %
Rail	71	76	0.4 %
Other	69	125	3.0 %
Total	2,186	3,082	1.7 %

The wide consensus is that this development must be avoided due to several reasons: The earth's oil reserves are limited, CO^2 emissions are causing the climate change, harmful substances and noise emissions are putting people's health in danger, especially in metropolitan regions, and rising fuel costs, which result from rising crude oil prices[6], should be evaded due to economic reasons.

---

[4]   Cf. World Economic Forum (2011), p. 11.

[5]   Cf. IEA (2009), p. 73.

[6]   Information on the increase of crude oil price see amongst others IEA (2009), p. 63ff.

The tourism sector is affected by these developments because change of location is the major fundament of tourism. Tourists are problem causers and victims alike: They are responsible for resource consumption and climate change, and travel is made more difficult for them when transportation becomes more expensive or rules are made adding restrictions.

Two means of travels are of utmost importance in the tourism industry: The passenger car is by far the most important vehicle for holiday trips. In Germany trips lasting longer than five days 48% of travelers use their own car. With 37% the airplane is the second most important method of transport.[7] The airplane has been publically criticized and labeled as a "climate killer", because airplanes are responsible for an above average increase in energy consumption and exhaust emissions due to the above average growth of air traffic. Once more tourists are in general users of carriers in the destination, which they use for excursions and sightseeing.

If global warming is to be confined to 2° Celsius then an 85 % reduction of CO^2 emissions by 2050 against 2000 is necessary. The European legal framework under the act of the European Parliament EU regulation 443/2009 allows an EU fleet consumption of 95g/km[8]. Such ambitious goals are only attainable if the basic structure of the transportation sector changes. Efforts to help electric vehicles (EVs) make a market breakthrough are currently at the forefront.[9]

Electro mobility seems to be a renaissance of the mobility concept: Few are aware that at the beginning of the 20th century automobile pioneers experimented with electric motors as well as combustion engines. Gustave Trouvé presented a vehicle with electric drive in Paris five years before Carl Benz applied for a patent for his vehicle.[10] Around the turn of the 20th century many cars in Hamburg, Berlin and New York were powered by electricity. Electric drive was very popular then, because cars with combustion engines were loud, polluted the air and had to be started using a hand crank. Hybrid drive was also known over 100 years ago: Ferdinand Porsche presented a Lohner-Porsche with an e-motor and combustion engine at a world exhibition in Paris in 1900. Ultimately, internal combustion engines (ICE) won the race on account of the travelling radius they provide and the lower-cost fuel they use.[11]

The current challenge is to reshape the global transportation system in terms of sustainability. Sustainability is given when intact ecological, social and economic structure are handed down to subsequent generations.

---

[7] Cf. FUR (2011), p. 4.

[8] Cf. EU (2009).

[9] Cf. Malomy (2011). McKinsey calculates that 68-93% of all traffic must be powered by electricity so that the target of an increase in temperature of 2°C due to global warming can be met.

[10] Cf. Mobility today (Mobilität heute) (2010), p. 5.

[11] Cf. Greenpeace (2011), p. 24, 29.

3   Determining "Sustainable Mobility"

"Sustainable mobility" is not defined as a universally understandable term. The German government has a loose understanding of the term when it comes to classifying sustainable mobility into three segments:[12] Efficient and alternative drives or fuels, diversion of traffic and traffic avoidance.

The first segment – ensuring efficient and alternative drive systems or fuels – aims at cutting primary energy consumption and consequently also at reducing pollutant emissions. The following mobility concepts are generally described as "sustainable".

- Conventional mobility concepts with efficiency improvements: Efforts being made by all modes of transport (road, air, rail and waterborne transport) to reduce fuel costs were reason enough to make varied approaches to optimize drives, so that the same traffic performance can be achieved using less fuel. However such optimized conventional mobility concepts cannot really be regarded as sustainable, as consumption of fossil fuels is further required and exhaust emissions are still being produced – though be it in less amounts.

- Electromobility concepts: These are all concepts which involve the drive being at least partly powered through the help of electric motors. Electric Vehicles (EVs), road vehicles with an electric motor, are the epitome of sustainable mobility. Electric drive concepts are not suited to air traffic due to the unfavorable performance/weight ratio. They also play a minor role in waterborne transport, whereas in rail traffic they are a given. Currently most efforts are being especially made to help LDVs with electric motors to make a market breakthrough. Mention is given here to hydrogen as well as fuel cell drives.

- Mobility concepts with alternative fuels: Alternative fuels are those which are completely or partly produced by renewable resources. These can be algae, maize, sugar cane or other plants. Such "biofuels" possess the advantage, that contrary to fossil fuels they grow back and produce less or no exhaust emission.

- Solar mobility concepts: Drive concepts which use solar power generated by solar cells attached to the vehicle are still in the early development phase. Here the values for primary energy consumption and exhaust emissions are zero. Remarkable pioneer projects can be found in air traffic with Solar Impulse[13], in shipping traffic with Planet Solar[14] and in road traffic with Solartaxi[15].

---

[12] Cf. Becker (2010).

[13] Cf. www.solarimpulse.com.

[14] Cf. www.planetsolar.org.

[15] Cf. www.solartaxi.com.

Changing means of transport can be sustainable if the change involves more re-source efficiency and causes less exhaust emissions. Politically intended for ex-ample is substituting parts of motorized private transport with short-distance pub-lic transport and the shift from air to rail traffic. Of course traffic avoidance is the most sustainable option. Here for example the substitution of business trips with video conferences is noteworthy.

4 Digression: Basics of Power Generation

The process of transport requires the input of energy.[16] Figure 1 shows the most important primary energy sources with their specific strengths and weaknesses.

Primary energy	Brief description	Strengths	Weaknesses
Crude oil	• Usage of heat energy through burning oil • Currently the most important energy source in the transportation sector	• Very high energy density • Very good deploy ability for means of transport • Optimal suitability for air traffic • Established crude oil distribution system	• Oil resources are coming to an end • Supply insecurities due to political uncertainties in oil producing countries • High pollutant emissions
Natural gas	• Usage of heat energy through gas combustion • Important energy source for generating electricity • So far little use in the transportation sector	• Natural gas reserves will outlast oil reserves • Switching quickly to and from gas power plants provides an ideal energy buffer • Compared to oil: Lower pollutant emissions • (Non-hydrocarbon methane)	• Natural gas reserves are coming to an end • High pollutant emissions (Hydrocarbons) • High infringement of the environment due to production and transport through pipelines
Coal (black coal and lignite)	• Usage of heat energy through burning coal • Important source of energy for generating electricity • No longer used in the transportation sector (earlier use by steam engines)	• Has probably the longest availability worldwide of all fossil fuels • Coal resources in countries which don't produce gas or oil (e.g. Germany)	• Extremely high pollutant emissions, especially lignite • High infringement of the environment, especially lignite production • Not suitable for any means of transport
Bituminous sands	• Currently of little importance as an energy resource	• Found in politically stable countries (e.g. Canada)	• Complicated mining and high production costs • High infringement of the environment • Low energy density
Nuclear power	• Power generation through nuclear fusion • Common source of energy for generating electricity • Not yet used in the transportation sector (exception: nuclear submarines)	• No toxic emissions • Low costs, if follow-up costs caused by nuclear accidents are not considered	• Limited uranium resources • High safety risks • Low level of acceptance in many countries • Not suitable for any means of transport

Fig. 1. Important primary sources of energy with their specific strengths and weaknesses

---

[16] The application of man power (e.g. cycling or walking tours) is neglected in the follow-ing. Changing location in this way is a marginal phenomenon in the tourism industry.

Primary energy	Brief description	Strengths	Weaknesses
Wind power	• Use of wind-produced kinetic energy which is generated by transformers • Common renewable energy source • Increasingly important for power generation	• Inexhaustible energy source • No pollutant emissions • Favourable cost structure in windy areas	• Necessity to expand the power supply system • High infringement of the environment • High costs for offshore wind parks • Inconsistent power production • Not suitable for any means of transport (exceptions: ships, gliders)
Solar energy	• Solar cells generate electric power • Relatively rare source of renewable energy • Increasingly important for power generation	• Inexhaustible energy source • No pollutant emissions • High potential of cost reduction in photovoltaic	• Necessity to expand the power supply system • Solar cells require a lot of space • High costs in regions with little sunlight • Inconsistent power generation in regions with little sunlight • Very limited use for any means of transport
Biomass	• Electricity generation through combustion of biomass (wood pellets, straw, grain, etc.) • Converting organic raw materials into oil • Still a relatively rare source of renewable energy • Increasingly important for power generation	• Renewable source of energy • Can be used to power modes of transport (in the case of bio fuels) • Bio fuels have similar qualities to fossil oils • Low toxic emission	• Release of greenhouse gases (which would be released anyway due to rotting) • Potential competition with food production • Water consumption • Deforestation
Geothermal energy	• Electricity generation through the Earth's heat	• Inexhaustible energy source • No pollutant emissions	• Not suitable for any means of transport
Hydropower	• Use of kinetic energy created by waterfalls (gravitational energy) generated by transformers • Currently of little importance • Increasingly important for power generation	• No pollutant emissions • Potentially low costs • Inexhaustible source of energy	• Very dependent on the natural environment • Infringement of the environment • Not suitable for any means of transport

Fig. 1 (continued)

In the first half of 2011 20.8% of kilowatt hours produced in Germany were generated by renewable energy.[17] In the course of this 7.5% was created by wind energy, 5.6% by biomass, 3.5% by solar energy, 3.3% by hydropower and 0.8% by others (electricity generated by waste-fuelled power station and other renewable energy sources).[18] The German government's aim is to increase the share of renewable energy up to 35% by 2020, to 50% by 2030 and to 80% by 2050.

On a global basis the consumption of primary energy will rise each year by 1.5% between 2007 and 2030 which corresponds to an increase of 40% during this

---

[17] With regard to conventional sources of energy in 2010 22% came from nuclear power, 19% from black coal, 24% from lignite and 14% from natural gas.

[18] Cf. German Federal Association for Energy and Water Management (2011).

period. In global terms the renewable energy's share of electricity generation will increase from 18% in 2007 to 22% in 2030.[19]

5 Concepts of Sustainable Mobility

Concepts of sustainable mobility can be found with all modes of transport. In the following the modes of transport, road, air and water will be looked at in detail. The mode of transport "rail" will be excluded, as this has long been considered sustainable. Therefore, radical changes towards sustainability, as can be currently clearly observed for other modes of transport, are not found for the mode of transport "rail".

5.1 Road Traffic

Efforts to increase efficiency of conventional combustion engines, have characterized activities for decades in the automotive industry. Fuel consumption is one of the most important decision criteria when buying a car.

Even though in the last decades clear progress has been made in terms of increasing efficiency, significant amounts of potential still seem to exist: Once again in the summer of 2011 the largest European energy saving race took place on the Lausitzring in Brandenburg which is carried out by the Shell Group. In 2010 a French team from Nantes managed to cover a distance of 4,896 km using just one liter of fuel. This equates to a fuel consumption of 0.02 liters per 100 km. In the category "Roadworthy Vehicles" a vehicle from the Trier University of Applied Sciences managed to travel 233 km using just 1 kWh of electricity in 2011.[20] These experiments show what efficiency reserves still remain. The German automotive industry believes that increasing efficiency by 25% by the end of this decade is possible.[21]

There are numerous types of fuel available to power vehicles. Table 3 outlines the amount of greenhouse gas emissions of different fossil and **alternative fuels**.

Most newly registered cars in Germany still have conventional drives, see Table 4. It is astonishing that only 541 electric vehicles were registered.

The main concept of sustainable mobility, which currently dominates economic and political discussions, is to ensure that electric vehicles ("electromobility")

---

[19] Cf. IEA (2009), p. 73.

[20] Cf. Shell (2011).

[21] Cf. Wissmann (2011), p. 2.

Table 3. Greenhouse gas emissions of different fuels. Source: Dena (2010), p. 10.

Fossil Fuels		Biofuels		Electrical drives	
Fuel	Greenhouse gas emissions in g/km*	Fuel	Greenhouse gas emissions in g/km*	Fuel	Greenhouse gas emissions in g/km*
Petrol	164	Natural gas with 20 % Biomethane	100	Hydrogen (EU mix)	174
Diesel (with particle filter)	156	100 % Bio-methane (bi-fuel, dung)	5	Hydrogen (100 % wind power)	8
Liquefied petroleum gas (LPG)	141	Ethanol (composition sugar beet, added waste material)	111	E-Mobility (EU power mix)	75
Compressed natural gas (EU natural gas mix)	124	Biodiesel (Composition: rape, added glycerin)	95	E-Mobilität (100 % power mix)	5

*Greenhouse gas emissions WTW in gCO^2eq/km. WTW: From the raw material's source (well) over the tank (tank) to the drive (wheel). gCO^2eq: CO^2 equivalent.

Table 4. Newly registered cars market share in terms of fuel type, Germany 2010

Type of fuel	Market share	Type of fuel	Market share
Petrol	57.30 %	Natural gas	0.170 %
Diesel	41.90 %	Electricity	0.019 %
Liquefied gas	0.28 %	Hybrid	0.370 %

Source: German Federal Department of Motor Vehicles, Flensburg, quoted in Gneuss (2011), p. 4.

make up a significant part of personal mobility. By way of example the German government has set a target of one million EVs to drive upon German roads by 2020.[22] The categories of vehicles reach from Pedelecs and Scooters over LDVs to busses for short-distance public transport, with an emphasis on LDVs.

Electric vehicles are said to have the advantage that they consume less primary energy (especially oil) and therefore cause lower exhaust emissions. Hereby the climate change problem is eased, health risk of people living in congested areas is

[22] The German government is nowhere near reaching this objective in the first quarter of 2011: Five hundred and eighty-two electric vehicles were registered which makes up just 0.08% of all registrations. Out of 42 million cars, which are on the road in Germany only 2,800 use an electric motor. See also the initiatives of the German Federal Association for E-Mobility (registered association) under www.bem-ev.de.

lowered and the dependence on oil is reduced. However if EVs are to actually produce less exhaust emissions and have a lower primary energy consumption, then power mainly coming from renewable energy sources must be used. If electricity would originate from old coal power stations, then "clean" vehicles would be running on "dirty" electricity. This is also a reason why car manufacturers are investing in green electricity projects.[23]

Electric vehicles are currently powered by several drive concepts which differ as follows:

- Hybrid electric vehicles (HEV): Without connection to mains with combustion engine plus electric motor.

- Plug-in hybrid electric vehicles (PHEV): With partial connection to mains and a combustion engine plus electric motor with batteries rechargeable using mains.

- Battery electric vehicles (BEV): Connection to mains necessary, with an electric motor, rechargeable batteries using mains and regenerative braking (recuperating breaking energy to recharge the battery).

- Fuel cell electric vehicles (FCEV): Without mains connection, chemical energy generation which powers the electric motor.[24]

- Range extender electric vehicles (REEV): After battery depletion a small combustion engine is used to increase cruising range.

Furthermore modifications of the concepts named above exist. The mobility concept of Better Place should get a mention here, where empty batteries can be swapped for full ones within ten minutes at service stations. Thereby problems related to battery charging time are reduced as well as insufficient cruise range of EVs. The company Better Place has entered into a cooperation with Renault-Nissan.[25]

Apart from the already mentioned positive attributes pure electric vehicles also demonstrate a number of negative features which hinder a quick market breakthrough:

- Range: To date most EVs available on the market have a low cruising range, on average about 100–150 km. Some vehicles have an even shorter cruise range. Cruise range such as the Tesla Roadster provides with 350 km is an extremely rare exception. Low and high temperatures, air conditioning and heating also reduce cruise range. Due to the short cruising range, EVs are barely suitable for overland travel like holiday trips. However cruising range seems to suffice for urban traffic or commuting because 80% of all car

---

[23] Cf. Handelsblatt, 29.08.2011.

[24] Cf. amongst others Becks (2010), p. 14.

[25] Cf.www.betterplace.com.

journeys are less than 70 km (in the case of Germany).[26] Car purchasers still feel that short cruising range is a major deficiency.

- Acquisition costs: Purchase costs of EVs are almost always over € 30,000, even for the smaller vehicles. The steep price primarily results from high costs of batteries (about € 10,000 – € 15,000), which are mostly lithium-ion batteries.[27] An EV doesn't pay off in the foreseeable future, despite having lower electricity consumption costs compared to petrol or diesel. The Mitsubishi i-MiEV for example has an average consumption of 16.6 kWh. This equates to electricity costs of about € 3.60 per 100 km for an electricity rate of 0.245 €/kWh.[28] Consequently numerous monetary and non-monetary government incentives are being discussed, with which car purchasers are to be encouraged to buy an EV. Car manufactures are also making an effort to release automotive purchasers from the burden of high battery acquisition costs by offering rental schemes for batteries.

- Availability of charging stations: Just like conventional vehicles which need filling stations, charging stations are essential for EVs. Domestic charging stations in one's own garage are especially suitable.[29] If EV drivers do not own a garage charging station nearby the home must be set up. Additionally charging stations can be set up in shopping centers, at the workplace or similar locations where people spend a lot of time. A low network density is a serious stumbling block for purchasing EVs.[30] Here a lot of investing needs to be done to achieve a satisfactory network density. A high network density seems to be a necessity for electro-mobility to break through the market. Incidentally charging stations must be equipped with payment facilities.

- Charging time: Charging time is largely dependent on the utilized voltage. Charging time can span from 30 minutes (for quick charging with high voltage) to about 6–8 hours when using domestic electricity which has 220V.[31] Charging technologies are being currently developed which enable

---

[26] Cf. EON (not dated), p. 17. See also Lubbadeh/Niemann (2011), p. 26. Growth from Knowledge (GfK) even found out asking 6.200 people in an empirical study, that in Germany's case 76% of all car journeys are shorter than 20 km, cf. GfK (2011).

[27] Raw materials such as lithium and rare earths will in the future drive costs of battery production; cf. Lubbadeh/Niemann (2011), p. 27. China possesses a high concentration of these raw materials from which China profits in terms of international competition.

[28] Cf. Frankfurter Allgemeine Zeitung (FAZ) from 12.04.2011, p. T3.

[29] Ideally the roof of the garage is covered with solar panels which generate electricity for the EV.

[30] See amongst others Bulczak (2011), p. 6. McKinsey have a different opinion on the matter, which believe that a dense network of public charging stations is not needed, cf. McKinsey (2011).

[31] Cf. Lubbadeh/Niemann (2011), p. 28.

electric cars to recharge batteries for about 50 km within ten minutes using quick charging.[32] The Better Place concept is something special which allows empty batteries to be exchanged for full ones at service stations within ten minutes. Batteries are not acquired with the purchase of a car. They are only lent and solely the charging process has to be paid for.

- Type of energy generation: Compared with conventional (combustion) engines EVs aim to significantly reduce exhaust emissions. This only works if the electricity used comes from renewable energy sources. If electricity originates from conventional energy sources (such as coal power stations), then pollutant emissions may even surpass those caused by conventional combustion engines.[33] If electricity came from the German network in 2011, then CO^2 emission would be 71g/km compared to 103 g/km of conventional combustion engines.[34] Therefore it must be ensured that electricity used by EVs comes from renewable energy sources.

- Network expansion: For a large use of EVs the problem arises that electricity consumers have to charge their EVs in locations other than those where electricity is generated by renewable energy sources. For example energy is produced by wind power plants near the coast and actual electricity consumption occurs in Munich. It is necessary to expand existing overland power supply networks appropriately which also increases electricity costs and is often opposed to by the population.

- Safety: Although lithium-ion batteries are nowadays seen as safe, fears still exist from car drivers, that in the case of an accident serious personal injury occurs.

- Standardizing plugs, data protocols and payment systems: A standardization of plugs, data protocols and the payment system is a must should EVs manage a market breakthrough. Car drivers wouldn't accept if plugs or payment systems are not compatible at various charging stations. Looking at the tourism sector it must be taken into consideration that vehicles in holiday traffic normally cross borders. Therefore not only national but also international normative regulations must be established.

In September 2011 the EV offer on the German market was still quite lucid: Important EV models are the Mitsubishi i-MiEV, the identically constructed LDV Citroen C-Zero and Peugeot iOn (leasing only) or the Tesla Roadster. Almost all car manufactures worth naming are planning an EV market launch in the next few years: Volkswagen (VW Golf blue e-motion, VW e-Up), Audi (R8 e-tron), Daimler

---

[32] Cf. EON (not dated), p. 17.

[33] Cf. the calculations in Lubbadeh/Niemann (2011), p. 29.

[34] Cf. Mobilität heute (2010), p. 5, which quotes an ADAC (German Automobile Association) study.

(E-Cell, Smart ed[35]), BMW (Megacity Vehicle and i3 and i8), Opel (Ampera, same construction as the dem Chevrolet Volt), Ford (Focus Elektro), Fiat (500 ev), Renault (Fluence Z.E., Kangoo Z.E., Twizzy Z.E., Zoe Z.E.), Volvo (C30 BEV), Saab, Toyota (EV2)[36], Honda, Mazda, Nissan (Leaf), as well as the Chinese provider BYD E6.[37] Most EVs are small – medium sized LDVs. Exceptions are for example the sports car Audi e-tron or the Tesla Roadster. However the number of EVs currently produced is still very limited.[38]

There are many technical differences between EVs and conventional vehicles with combustion engines. The drive unit consists of an electric motor instead of a combustion engine and the transmission is left out, which must be replaced by extremely high-performance batteries.[39] Conventional filling stations are becoming redundant, instead charging stations are required. It should also be mentioned that due to their batteries, EVs can function as temporary mobile storage for electric energy which originates from a power mix with a growing share of renewable energy and therefore can provide a contribution towards solving volatility issues this type of energy has. Traditional providers such as engine manufacturers and mineral oil companies are losing importance whereas electronics and electricity companies as well as battery producers are gaining significance. Today the electro-technical share of value added per car is approx. 35%, for EVs this share increases to about 70%.[40] New materials are gaining importance. EVs must be significantly lighter than conventional vehicles.[41] As a result the entire value-adding chain in the automobile industry will be rearranged.[42] It is currently not foreseeable which companies out of which regions of the world will emerge as winners and which as losers due to this rearrangement. Fears still exist that automobile manufacturers will lose importance and that emerging countries like China (who are considered to be very ad-

---

[35] Daimler conducted a study with the chemical firm BASF where transparent solar cells on the roof of the Smart generate additional energy. The vehicle is also clearly lighter due to modern plastics being used.

[36] Toyota must be given the merit for introducing a series hybrid vehicle (Toyota Prius) into the world market early on. This results in Toyota having a pioneer income return.

[37] Cf. Neue Mobilität (2011), p. 24 and own analyses carried out by IAA 2011.

[38] In 2010 6,665 Leafs (Renault-Nissan), 3,917 i-MiEVs (Mitsubishi), 3,444 iOns und C-Zero (PSA), 2,881 QQ3s (Chery Auto), 1,984 Think Citys (Think), 1,878 Volts (GM), 1,844 Midsize Sedan Codas, 1,224 e6 (BYD), 800 Smarts ed (Daimler) and 769 Tesla Roadsters were produced, cf. PwC, quoted in: FAZ from 06.01.2011.

[39] A conventional combustion engine of a LDV has 1,400 parts installed in the powertrain where an electric motor only has 210.

[40] Cf. Schneider (2010).

[41] Metals used by modern automobile manufacturers are being increasingly substituted with plastics. A good example is the automobile manufacturer BMWs utilization of carbon. BMW secured its resource access by holding a share in SGL Carbon.

[42] Also cf. Arthur D. Little (2010).

vanced in terms of battery technology and production) will gain significance compared with traditional producing countries such as Germany.

A special role is given to **solar vehicles**. Here electricity is generated by solar cells attached to the vehicle to power electric motors. Pioneering work was carried out by people who demonstrate suitability for everyday use using prototypes. The Swiss Louis Palmer already managed to travel around the world in 2007 and 2008 with his self-constructed vehicle which had solar cells installed on a trailer.[43]

The introduction of electromobility means no less than a change of system for all parties involved.[44] The change of the model range of automobile manufacturers will lead to a fundamental transformation of the provider structure in the automobile industry. Figure 2 shows the "Eco-system of electromobility" including fields of activity and challenges.

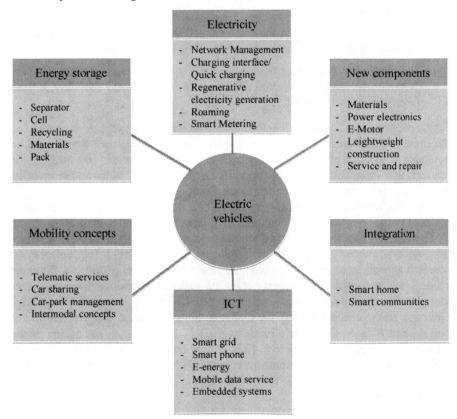

Fig. 2. Eco-system of electromobility

Source: IAA (2011).

---

[43] See www.solartaxi.com.

[44] Cf. Hornig/Kasserra (2011), p. 80.

Recently formed value-added partnerships prove changes in the industry structure:

- General Motors and the Korean electronics group LG are planning a joint development of electric vehicles.
- Volvo and Siemens are planning an extensive cooperation whereby Siemens produces electric motors for the Volvo C 30 Electric. As Volvo belongs to the Chinese automobile group Geely, Siemens could gain access to the huge Chinese market. [45]

Cooperations of German automobile manufacturers are not as extensive:

- Even though Volkswagen buys the batteries for the small passenger car VW Up from the electronics group Sanyo, the Japanese do not participate in the vehicle's development.
- Daimler cooperates with the chemical group Evonik for the production of batteries but reserves the construction of the Elektro-Smart for itself.
- Although BMW has its batteries for the city electric car i3 delivered by the German-Korean joint venture SB Limotive of Bosch and Samsung, it does everything else by itself.[46]

Rental car providers are expanding their fleet with electric vehicles. Avis announced the order of 500 Renault Fluence Z.E. and Kangoo Z.E., which are supposed to be bookable at the end of 2011. Sixt is planning to order just under 100 Stromos for business and local clients from the provider German E Cars. More and more often electric cars are being added to company car fleets.

As a market breakthrough is very unrealistic without government incentives, national governments in many countries have launched incentive measures.

The German government formed an initiative called NPE (nation platform electromobility), where different sectors of industry, on all political levels, science and research as well as consumer and environmental associations contribute. Concrete proposals how to reach the targets set by the national development plan electromobility are compiled in seven work groups. Ultimately the goal to establish Germany as the leading market for electromobility has to be reached and to put one million EVs on Germany's roads by 2020.[47] According to the notion of the NPE research funds amounting to € 4 billion will be required of which the government is to provide an as yet unknown amount. In its second final report the NPE presents a package of measures to compensate the cost gap and stimulate demand for electric vehicles:[48]

---

[45] China has for a long time made good experiences with electric scooters: in 2011 about 150 million electric scooters are driving in China's roads. With this in mind the sales volume of electric vehicles will increase quicker in China than in Europe.

[46] Cf. FAZ from 27.08.2011.

[47] Cf. The German government (2010).

[48] Cf. NPE (2011), p. 7.

- EV preference for parking, permission to use bus lanes as well as supporting new, intelligent car sharing concepts.

- Compensation for disadvantages of private use of company cars with electric drive.

- Special amortization for commercial purchase of EVs.

- Loans at reduced interest rates from KfW (German government-owned development bank) for the private purchase of EVs.

- Granting yearly tax incentives, based on the storage capacity of an EV.

In May 2011 Germany increased the research support up to 2012 by € 1 billion. So far the following projects were supported: Electromobility in public space, electrochemistry research, information technology for electromobility, lithium-ion-batteries, system research for electromobility, energy research, traffic research, field experiments in LDV traffic, fleet experiments in trade traffic, battery test center.[49] So far € 200 million have been provided for "showcase projects". From September 2011 onwards consortiums consisting of automobile producers, electricity companies, short-distance public transport companies and communities from presumably eight different regions in Germany will apply.[50] Funds totaling € 200 million is supposed to be topped up to € 600 million through own means.

Previous regional government efforts to support electromobility have had different success. According to the Electric Vehicle Index EVI from McKinsey Germany is currently positioned midfield with a value of 30 along with Japan and China. USA and France are positioned at a value of a solid 40; Italy, South Korea, Great Britain and Spain are positioned at a value under 25.[51]

It can be established that Germany's support measures are minor when compared internationally. For example in other countries it is common that direct purchase incentives come in the form of subsidies when buying a car. "In many European countries, such as France, Great Britain, Spain and Portugal the purchase of an EV gets directly supported by an average premium payment amounting to € 5,000."[52] The Chinese offer substantial support measures. Up to 2020 funds amounting to € 11 billion will be provided to develop electromobility. Whether the development of electromobilty is successful remains to be seen. Distinct preferences of Chinese car purchasers towards conventional cars are sometimes reported.

According to an up-to-date study carried out by the GfK, 10% aged over 18 in Germany expressed their intention to buy an EV in the next few years. In the UK

---

[49] Cf. FAZ from 16.08.2011.

[50] Cf. Wirtschaftswoche (2011), p. 52ff.

[51] EVI measures by how many percent electromobility reaches a country, which is predicted by experts for 2020, cf. Wirtschaftswoche (2011), p. 54.

[52] Bachmann/Mayer (2011), p. 90, also see Gneuss (2011), p. 4.

it was 8.7%, in Netherlands 29%. Fifty-six percent of people asked in Germany are also prepared to pay a higher price, whereby the maximum surcharge may only amount to € 2,000. Most of the people questioned will know that surcharges will be higher which leads to the purchase price being an important stumbling block. Yet the greatest obstacle for buying an EV is the extremely limited cruising range of the vehicle. A cruising range less than 150 km is only satisfactory for 6.6 % of German, 26.2 % of British and 13.4 % of Dutch people. A third important obstacle for purchasing an EV is the limited availability of charging stations.[53]

Mobility providers such as Deutsche Bahn, Lufthansa, car rental services such as Sixt etc. have an important role to play when it comes to expanding electromobility: They can provide EVs in destinations and thereby introduce wealthy clients and opinion leaders to EVs.[54]

Car sharing refers to mobility concepts, where several people share an undetermined LDV and do without the possession of a LDV. The LDVs are property of a company which provides members in a certain city with the use of LDVs. To be able to use LDVs users pay a fee that is mostly based on duration of use and kilometers driven. At present there are a number of car sharing providers in Germany and other countries.[55]

Car sharing concepts have several positive effects on the environment:[56] On the one hand the amount of cars produced and used will decrease due to many people sharing a car and on the other hand car sharing providers often employ EVs which drive on certified green electricity. For example the cooperation between Cambio and Greenpeace-Energy in Hamburg offers EVs, namely the Mitsubishi i-MiEV. Car2Go which involves the automobile manufacturer Daimler exists in Vancouver, Austin-Texas, Hamburg and Ulm. Car2Go, for example, also offers Elektro-Smarts in the municipal area of Ulm. The city Ulm is setting up a charging infrastructure consisting of 24 charging stations. This business model is currently running through practical trials where technical details and consumer behavior are tested. The Car2Go business model could set an example for other classical car rental providers in busy city destinations such as London, Paris, Rome or Berlin. Registered (regular) customers of a car rental provider could use the environmental friendly vehicle fleet in the urban area of a city destination without having to check-in first.

Concepts which include modal shifts in transport to more resource efficient means of transport, such as bus and rail, count as very sustainable. Especially busses demonstrate a very low consumption of primary energy and low production

---

[53] Cf. GfK (2011), p. 90 see also Future Foundation (2010).

[54] Cf. also Arthur D. Little (2010).

[55] For example see www.book-n-drive.de, www.car2go.com, www.stadtmobil.de or www.flinkster.de.

[56] However sometimes it is criticized that car sharing can lead to people using short-distance public transport less, as cars can be used at any time for little money without having to own a car.

of pollutant emissions. A unique form of modal shifting is performed by companies with a business model that allows people to express their transportation needs on internet platforms. Should enough requests come together for certain routes then busses are rented to cover these[57]. These business models could further spread due to the extensive presence of modern information technology and high internet usage.

5.2   Air

The air traffic's share of greenhouse gas emissions caused by people is a good 2%. This share would increase due to the expected annual growth of world air traffic of 4–5% so long as the air traffic industry doesn't introduce any suitable counter-measures. The industry committed itself to ensure a neutral CO^2 growth of air traffic starting 2020 and from 2050 onwards greenhouse gas emissions are to be halved compared to 2005.

Sustainable mobility measures in air traffic are primarily directed at increasing **fuel efficiency**. Air traffic too shows a direct correlation between fuel consumption and CO^2 emissions: Saving 1 kg of kerosene leads to a reduction of CO^2 emissions by 3.16 kg. Airlines have massive interest anyhow in reducing kerosene consumption, as fuel costs make up 30% of operating costs of IATA airlines in 2011. Fuel costs of all IATA airlines totaled to about US $ 176 billion.[58] Substantial progress has been made in the past decades in terms of fuel efficiency. Nowadays new aircrafts consume about 70% less fuel than over 40 years ago. The last ten years alone increased fuel efficiency by 20%. Modern aircraft consume about 3.5 liters per 100 passenger kilometers, the A380 and die B787 about 3.0 liters per 100 passenger kilometers. The aim is to further increase fuel efficiency by 25% by 2020.[59]

Along with aircraft and engine manufacturers, airlines are also making efforts in the operations department. Route optimizations are targeted here (e.g. non-stop flights) to reduce fuel consumption by varying cruse speeds and optimizing air-craft load.[60] The airport segment is making efforts to optimize ground traffic flow and starting/landing processes. In addition the air traffic management sector is supposed to reduce jet fuel consumption and CO^2 emissions through route naviga-tion optimization and avoidance of holding patterns.[61]

A very promising approach to reduce CO^2 emissions is to use alternative fuels. A 3% admixture of alternative fuels reduces CO^2 emissions by 2% – this currently equates to an annual reduction of ten million tons of CO^2. Alternative fuels consist of regenerative, biological raw materials such as algae, babassu palm oil, camelina, halophytes, jatropha and switchgrass (so called "biofuels").

---

[57] For example see internet bus platforms such as www.deinbus.de.

[58] Cf. IATA (2011): Assumption: US $ 110 per barrel crude oil.

[59] Cf. IATA (2011).

[60] Also cf. the environmental strategy 2020 of the Deutsche Lufthansa AG, under: www.lufthansa.com.

[61] See European Commission (2011). Also see Sterzenbach/Conrady/Fichert (2009), p. 69ff.

The air traffic sector demands the following from biofuels:

- Biofuels must be able to be mixed with conventional kerosene made from fossil fuels and must be able to use the same supply infrastructure.

- They must demonstrate the same technical specifications as conventional fuels. Therefore biofuels must sustain low temperatures of $-40°$ or $-47°$ Celsius (corresponds to Jet A – and Jet A-1-Fuel) and turbine heat as well as have a high energy density (Jet A-Fuel has an energy density of 42.8 MJ/kg).

- Biofuel isn't allowed to be in competition with the food chain, to cause fresh water competition, is neither allowed to promote deforestation of pasture lands or put the biodiversity in danger.[62]

IATA believes that an admixture of 6% biofuel is achievable by 2020. This would mean a 4% reduction of CO^2 emissions. In its environmental strategy 2020 Lufthansa aims to add up to 10% regenerative fuels by 2020.

So far a number of experiments using biofuel have been carried out: In 2008 Airbus flew with an A380 with a GTL engine using the Fischer Tropsch-method, on 12.10.2009 Qatar Airways flew using a 50% GTL admixture in all four engines of an A340-600.[63] United flew with a A319 using 40% GTL. Between 2008 and 2011 a total of seven airlines flew using biofuels: Virgin Atlantic, Air New Zealand, Continental, JAL, KLM, TAM and Interjet. Overall the results are very promising: No airplane or engine adjustments had to be made, the addition to conventional fuel jet was unproblematic and fuel efficiency didn't drop. Since 15.07.2011 Lufthansa uses a 50% biofuel admixture in one of the A321s engines for the Hamburg-Frankfurt flight route to examine long-term biofuel effects. The initiative "Aviation Initiative for Renewable Energy in Germany" (AIREG) was founded in Germany to encourage the use of regenerative energy sources in air traffic over Germany.[64]

In Europe, air traffic is to be included in the Emission Trading Scheme (ETS) in 2012. ETS plans that every airline which takes off and/or lands in Europe has to purchase emission certificates. This would achieve internalization of external effects: Those responsible for pollution would take the blame for the effects caused. It can be expected that the ETS would incent the use of fuel-saving airplanes. However it is also to be expected that unfair competition in global air traffic would arise as for example airlines from the Gulf States would be less or even not burdened at all by the ETS. This is where European airlines begin to criticize the ETS.

For the next decade **electromobility concepts** in air traffic are still not technically feasible. The main problem is the power to weight ratio which reaches its

[62] Cf. IATA (2011).
[63] GTL-fuel produces lower pollutant emissions than conventional jet fuel.
[64] Cf. www.aireg.de.

physical limits. Conventional fuels such as kerosene have an energy density which batteries don't even come close to. Nowadays 30 kg of the best batteries are needed to store the same amount of energy as in one kilogram kerosene. [65] Crude oil counts as an excellent energy source in air traffic. Within the air traffic industry there is a saying: "The last drop of crude oil will be used by airplanes".

A far more observed pioneer project, which however is irrelevant for air traffic in the foreseeable future, is the **Solar Impulse** project of the flight pioneers Bertrand Piccard and André Borschberg. The Solar Impulse project aims to travel around the world using a solar airplane. The solar airplane currently used has four propeller motors which are powered only by electricity. The electricity is generated by solar cells installed onto the wings and stored in batteries for night flights. On 08.07.2010 the solar plane managed to fly 26 hours straight from the Swiss airport Payerne. Solar energy stored during the daytime flight was used at night. [66]

Airports are making a special effort to reduce burdens for people and the environment to improve the acceptance of air traffic from surrounding regions. Airport measures are primarily directed at reducing noise exposure, CO^2 emissions and use of space. [67]

Intermodal traffic concepts are suitable to reduce environmental pollution. The AIRail concept of Deutsche Lufthansa AG lightens the arrival by using railway connections to Frankfurt from Cologne and Stuttgart which leads to a modal shift to (environmental friendlier) rail traffic. Passengers can already check-in at Cologne or Stuttgart main station, i.e. they receive their boarding card and seating details whilst also being able to check-in luggage.

5.3   Water

It is little known that global shipping is responsible for approx. the same amount of pollutant emissions as global air traffic. In passenger traffic the cruise segment enjoys special attention, as this segment has for many years been developing faster than any other tourism segment worldwide. This is why environmental impacts caused by cruises have become more noticed. Air pollution (ship CO^2 emissions), water pollution, noise exposure, ship vibrations along with infringement on the environment by building ports and piers have become the subjects of discussion.

The branch seems to have understood the problem and is trying to contain harmful environmental impacts. [68] Success is actively communicated in Sustainability Reports. For example take a look at the in Figure 3 presented extract from the RCCL 2009 Sustainability Report:

---

[65]   Borschberg (2011).

[66]   See www.solarimpulse.com.

[67]   Cf. Beisel (2011).

[68]   Cf. detailed: ECC (not dated), p. 44ff.

- Fuel consumption down 3.7% – almost double the 2% target reduction
- Total waste brought ashore down from 1.5lb per available passenger capacity day in 2008 to 1.4lb
- Hazardous waste reduced by 25%
- Chemwatch database and Green Rating System launched on all ships
- Advanced Wastewater Purification systems – beyond compliance standards – installed fleetwide at cost of more than US$150 million
- Ocean Fund awarded US$484,000 to 14 marine conservation and environmental organizations.

Fig. 3. RCCL sustainability report extract

Source: ECC (not dated), p. 53

With regard to ship drives and engine fuel consumption it should be mentioned that for many years **solar drives**, which are completely free of emission, are used in niche markets. This occurs especially on European waters in densely populated areas, on lakes or rivers, where powerful drives and speed don't play a major role. Instead, the focus here is on aspects such as water pollution and noise control as well as zero emissions.

The following are noteworthy examples: Since September 2009 the solar ferry Helio travels on the Lake Constance[69], since 2000 the Spree Shuttle drives on waters in the Berlin city zone.[70] In 2000 the "Alster Sonne" commenced service in Hamburg.[71] The solar catamaran "MobiCat", with a capacity for 150 passengers plies the lakes of Jura in Switzerland since 2001.[72] A solar ship drives on the Neckar River since 2004 with transportation capacity for 250 people.[73] The first crossing of the Atlantic with a solar ship, the "Sun21", followed in 2007. The Sun21 and its crew were honored with the Guinness World Record Award for the fasted crossing of the Atlantic using a solar ship.[74] Since 2007 the research and laboratory ship "Solgenie" of the Constance University drives on the Lake Constance using a hybrid drive which generates power by using fuel cells (along with solar cells).[75] Since 2009 the first solar-powered passenger ship, the Solar-aquabus "Solon C 60", on the waters of Germany's capital Berlin.[76] Since 2010 the

---

[69] See http://www.ecolup.info/docs/standard.asp?id=6350&domid=629&sp=D&m1=926&m2=590&m3=6347&m4=6350.

[70] See http://www.spree-shuttle.de/.

[71] See http://www.photon.de/news/news_technik_00-07_alstersonne.htm.

[72] See http://www.bielersee.ch/de/schiffsmiete/flotte/ems-mobicat.283.html.

[73] See http://www.hdsolarschiff.com/de/.

[74] See http://www.transatlantic21.org/de/.

[75] See http://www.htwg-konstanz.de/21-06-07-Bootstaufe-Solgeni.1584.0.html.

[76] See http://www.schiffskontor.de/cms/solon-solarschiff-berlin.html.

first stages of a round-the-world trip are being covered using the Solar-wave.[77]Nowadays solar water taxis and luxury solar yachts are being built.[78]

A lot of attention was also being paid to the round-the-world trip using MS TÛRANOR PlanetSolar. The construction of the world's largest solar boat MS TÛRANOR PlanetSolar is an event with huge symbolic power for the progress of solar navigation. With its approx. 500 m² solar panels the MS TÛRANOR PlanetSolar can also navigate up to three days without direct sunlight. The boat is supposed to symbolically carry the message of the possibilities of using renewable energy "around the world". The long-term use capability is to be tested during a round-the-world trip. The trip around the world was 80% complete in August 2011.[79]

Efforts also exist to achieve sustainability mobility in ocean freight traffic. The design study Orcelle incorporates the use of fuel cells, solar, wind and wave power to drive a car carrier.[80]

6   Sustainable Mobility in Tourism Destinations

Touristic destinations rely on capable traffic systems and an intact environment. Especially fast growing metropolises are already suffering under traffic problems and burdens of air and noise pollution today. EVs are ideal means of transport for heavily encumbered metropolises, as EVs don't produce any emissions and hardly any noise. If the majority of vehicles in the city were powered by electricity then districts near main roads would significantly gain in habitation quality and asthma, cardiovascular and cancer illnesses would become scarcer. Especially in fast growing mega-metropolises in Asia and Africa mistakes made by American and European cities could be avoided by making timely adjustments.[81]

Sustainable mobility approaches can already be found in several regions. In the Alpine region (Germany, Austria, Switzerland, Italy, France, Slovenia) 13 pilot projects closely linked to tourism are being carried out. Here EVs or biofuel powered means of transport are being tested in short-distance public passenger transport, solar boats are being used on alpine lakes and e-bikes and EVs are being rented.[82]

---

[77] See http://www.solarwave.at/autark-um-die-welt.html.

[78] See http://www.solarwaterworld.de/produkte/passenger/solar-uno-wassertaxi.html und http://www.solarwaterworld.de/produkte/yacht/suncat-46.html.

[79] See http://www.turanor.eu/ and http://www.planetsolar.org/de/home.html.

[80] „The E/S Orcelle represents our vision for zero-emission car carrying. The idea combines fuel cells, wind, solar and wave power to propel the vessel, that will need no oil or ballast water. A car-carrier like this will never be built in its entirety but we hope to see some of the elements in future generation of vessels.", in: http://www.2wglobal.com/ www/environment/orcelleGreenFlagship/index.jsp.

[81] Cf. Lubbadeh/Niemann (2011), p. 26.

[82] See www.co2neutralp.eu.

The project Alpmobil is worth mentioning, where tourists in the Swiss region of St. Gotthard can rent EVs from the brands Think and Twingo in different locations. Charging stations and parking spaces are provided free of charge.[83]

In Asia the project endemic to the Philippines for the use of "EJeepneys" (electrical drive minibuses for the transport of tourists and locals) is exemplary.[84] A test phase has been started in Singapore with Mitsubishi i-MiEV and e-Smarts. By 2012 63 charging stations will have been set up in the city zone.[85] Singapore is ideal as a region for EVs because the majority of trips made are under 40 km/day and 90% of vehicles are parked at home, at offices or shopping centers approx. 22 hours per day. "One advantage that we have is that we are a small compact urbanized environment, which makes it in a way, convenient. Your travelling distances are not too long. It's not difficult for you to set up charging stations around the island."[86] In USA, California seems to be developing into early adopter market of electromobility. San Francisco is making noticeable efforts to reduce CO^2 emissions in the city and has once again underlined its plans to grant fee charging of batteries at all city charging stations by 2013.[87]

7   Conclusion and Outlook

No topic is being as intensely discussed as electromobility in the transport sector. This seems to be ideal solution for sustainable mobility for government policy makers and providers in the transportation sector.

An electromobility traffic system is barely comparable to nowadays traffic systems. On the way to extensive electromobility a complete upheaval of the traffic system is unavoidable.

The traffic system changes are in many ways a "chicken-and-egg situation". For example on the one hand clients will barely be inclined to purchase EVs if a suitable charging infrastructure has not been extensively established. Yet on the other hand these can hardly be set up in an economically viable way should demand go missing.

An electromobility market breakthrough is only to be expected if framework conditions are put in place by political policy makers. These have to be decisive, consistent and coordinated internationally. A special responsibility is given to the automobile purchaser: Environmental protection can no longer be demanded to be

---

[83]   See www.alpmobil.de.

[84]   See www.ejeepney.org.

[85]   See www.channelnewsasia.com.

[86]   Chee Hong Tat (2011).

[87]   See http://www.smartplanet.com/blog/transportation/san-francisco-will-charge-your-electric-car-for-free-until-2013/380 and http://www.sfgate.com/cgi-bin/article.cgi?f=/c/a/2011/05/09/BAJM1JE0CJ.DTL.

free of charge, additionally cherished habits must be thought over and changed if necessary. Even if favorable framework conditions exist it can hardly be expected that a transformation of the traffic system will take place short-term as it involves a process which will take decades.[88]

References

Arthur D. Little (2010): Winning on the E-mobility Playing Field, in: www.adl.com/e-mobility

Bachmann, P./Mayer, C.A. (2011): Förderungen im europäischen Vergleich, in: Neue Mobilität, p. 90 f.

Becker, M. (2010): Elektromobilität – Die Strategie der Bundesregierung, Referat IG I 5 Umwelt und Verkehr, Bundesministerium für Umwelt, Naturschutz und Reaktorsicherheit, Speech from 4th May 2010.

Becks (2010): Was ist Elektromobilität? Eine Definition für den Wegweiser, in: Becks, T./ De Doncker, R./Karg, L. et al. (eds.): Wegweiser Elektromobilität, Berlin – Offenbach u.a. 2010, p. 14 – 17.

Beisel, R. (2011): Acceptance of Aviation in the Airport Environment: Best Practice Examples, in: Conrady, R./Buck, M. (Hrsg.): Trends and Issues in Global Tourism 2011, Heidelberg 2011, p. 51 – 57.

Borschberg, A. (2011): Speech at the ITB Eco-Mobility Day of ITB Berlin Convention, March 2011.

Bulczak, L. (2011): Tanken per Ampel-Stopp, in: E-Mobilität, Reflex-Verlag 2011, p. 6.

Bundesregierung (German government) (2010): http://www.bundesregierung.de/Content/ DE/Artikel/2010/05/2010-05-03-elektromobilitaet-erklaerung.html.

Chee Hong Tat (2011): Interview with channelnewsasia.com, in: http://www.channelnews-asia.com/stories/singaporelocalnews/view/1137133/1/.html (Note: Chee Hong Tat is the CEO of Singapurer Energy Market Authority).

Dena (2010): Erdgas und Biomethan im künftigen Kraftstoffmix, Berlin 2010.

ECC (o.J.): European Cruise Coucil 2010/2011 Report, Brussels, not dated.

EON (o.J.): Elektromobilität, Düsseldorf, not dated.

EU (2009): Regulation (EC) No 443/2009 of the European Parliament and of the Council of 23 April 2009 setting emission performance standards for new passenger cars as part of the Community's integrated approach to reduce CO2 emissions from light-duty vehicles.

European Commission (2011): in: http://ec.europa.eu/transport/air/environment/environment_en.htm

FUR (2011): The 41st Reiseanalyse RA 2011, in: www.fur.de.

Future Foundation (2010): Plugged-in report. How Consumers in the UK view electric cars, no location named, 2010.

GfK (2011): Sustainable Mobility – Insights on Customer requirements and Purchase Intentions in Germany/UK/NL, Presentation at the ITB Eco-Mobility Day, Berlin 2011.

Gneuss, M. (2011): Eine lange spannende Reise, in: E-Mobilität, Reflex-Verlag 2011, p. 4.

Greenpeace (2011): Greenpeace Magazin 2.11.

---

[88] Also cf. Heymann (2009).

Heymann, E. (2009): Automobilindustrie am Beginn einer Zeitenwende, Deutsche Bank Research, Frankfurt/Main 2009.

Hornig, C./Kasserra, S. (2011): Es wird nicht reichen, nur ein Elektroauto zu bauen, in: Neue Mobilität, p. 80 f.

IAA (2011): Various information and charts at the IAA 2011.

IATA (2011): In www.iata.org.

IEA (2009): World Energy Outlook, Paris 2009.

Lubbadeh, J./Niemann, C. (2011): Elektroautos können klimafreundlich sein, in: Greenpeace Magazin 2.11., p. 23 ff.

Malorny, C. (2011): Thesen zur Elektromobilität in Deutschland, Speech in Bad Gögging, July 2011.

McKinsey (2010): Press release: Neue McKinsey-Studie – Elektromobilität in Megastädten: Schon 2015 Marktanteile von bis zu 16 %, in: www.mckinsey.com.

Mobilität heute (2010): Media planet, no location named, 2011.

Neue Mobilität (2011): Das Magazin vom Bundesverband eMobilität, Jan. 2011.

NPE (2011): Zweiter Bericht der Nationalen Plattform Elektromobilität, Berlin 2011.

Progtrans (2011): World Transport Report 2010/2011.

Schneider, J. (2010): Technologiesprung zu einem neuen Mobilitätssystem, in: E-Mobility, Special edition der FAZ from 23./24.10.2010.

Shell (2011): in: http://www.shell.de/home/content/deu/aboutshell/our_commitment/eco_marathon/concept/

Sterzenbach, R./Conrady, R./Fichert, F. (2009): Luftverkehr – Betriebswirtschaftliches Lehr- und Handbuch, 4th ed., Munich 2009.

Wissmann, M. (2011): Grußwort des Verbandes der Autobilindustrie VDA, in: Automobilzulieferer, Special edition of the Association of the Automotive Industry, Sept. 2011, p. 2.

World Economic Forum (2011): Repowering Transport, Project White Paper, Geneva 2011.

List of Abbreviations

CAGR	Compound annual growth rate
Dena	Deutsche Energie-Agentur GmbH
ECC	European Cruise Council
FUR	Forschungsgemeinschaft Urlaub und Reisen e.V.
IAA	Internationale Automobil Show
IEA	International Energy Agency
LDV	Light duty vehicles
Mtoe	Million tons of oil equivalent
RCCL	Royal Caribbean Cruise Lines
VDA	Association of the Automotive Industry (Germany)
WEF	World Economic Forum

Eco-mobility – Will New Choices of Electric Drive Vehicles Change the Way We Travel?

Björn Dosch

1 Introduction

E-mobility is as topical as few other issues in traffic. While politicians praise its potential to add to more sustainability in travel and transport, manufacturers hope for new markets and products that will counter the widespread notion that cars cannot be the solution to contemporary problems, namely the conflict between traffic, the environment and scarce energy resources. Consequently, all major car manufacturers announce new models based on electric drive. Smaller companies offer fresh concepts and hope to break into saturated markets. Many national governments have set ambitious targets for electric drive vehicles (EVs) to be sold. E.g. Germany is heading for 1 million EVs in the national car fleet by 2020, the UK for 1.55 million and France even for 2 million.[1]

This article tries to show why and how electric drive vehicles are thought to contribute to "eco-mobility". Potential benefits in solving important drawbacks of today's mobility will be discussed in section 2.

In the general discussion the notion of electric drive vehicles mostly refers to passenger cars whose internal combustion engine (ICE) is amended or totally replaced by battery-electric power-train. This article is written from this perspective. Public transport, electric bicycles (or even heavy-duty trucks) will not be covered though the former both can have important significance in touristic travel and activities. Two major technologies are the basis for electric drive vehicles, battery electric vehicles (BEV) and plug-in hybrid electric vehicles (PHEV). The technical approaches to e-mobility will be outlined in section 3. Further on, other concepts of "eco-mobility" will be described briefly, since electric drive should not be considered as the sole path of automotive development.

New EV technologies will not come without immanent challenges. Batteries as can be built today offer limited capacity (i.e. driving range) and come at high costs. Those burdens for market success need to be considered when investigating the future of electric drive vehicles. They will be discussed in section 4.

---

[1] See International Energy Agency 2009, pp. 18–19 for an overview.

Before reaching some conclusions, section 5 will give a few forecasts about the role EVs can play in car sales. Since in all countries market development is in a very early stage with few cars to buy and little experience being available, the text will focus on the case of Germany as an example of the more saturated automotive markets in the industrial countries.

2   Benefits of Eco- or Electric Mobility

While transport is widely accepted as fundamental to our economy and society its sustainability is challenged. This is especially true for the passenger car as the backbone of passenger transport in Europe (accounting for more than 70 % of the overall passenger-kilometers traveled in the EU). In its new white paper the European Commission stresses that the oil dependence of Europe needs to be addressed and greenhouse gas emissions need to be reduced drastically with the goal of limiting climate change below 2°C. Among other approaches, new technologies for vehicles shall contribute to these objectives. Especially in urban transport where also air-quality is an important concern, a gradual phase-out of "conventionally-fueled" vehicles is called for, halving their numbers by 2030 and phasing them out completely by 2050.[2]

Thus, electric drive vehicles promise to address important issues in transport policy as well as reduce negative by-products of road transport:

- CO_2 emissions with their harmful contribution to the greenhouse effect
- dependence on scarce sources of fossil energy, mostly crude-oil
- toxic emissions and their impact on (local) air quality

2.1   Energy-Use and Carbon Footprint

Efficient use of energy is a primary criterion when assessing different drive technologies. Actually, electric motors feature considerable better power efficiency than the combustion engines used in cars nowadays. But this advantage diminishes when electricity generation and transmission are considered. When regarding not only the energy used to drive a car but also to produce it, ICE cars can even perform better than EVs. Though, for the carbon footprint of travel the quality of primary energy used is much more important than power efficiency. An EV with a consumption of 20 kWh/100 km will emit about 120 g CO_2/km based on the current German energy mix (ca. 600 g CO_2/kWh) – no better than fuel efficient ICE cars. When using electricity from other sources the result could be altered altogether, though. The same car would emit only 76 g CO_2/km based on the overall

---

[2]   See European Commission 2011.

energy mix of the European Union. Using low-carbon primary energy – probably wind, water or solar power – the carbon footprint of an EV can even be reduced drastically in comparison with an ICE car.[3]

This flexibility in using different sources of primary energies is a big advantage of EVs not only because of climate policy. It also allows for greater independence of fossil energy sources and of the countries exporting them. In 2008, mineral oil accounted for 38.7 % of final energy consumption in Germany. In transport the figure was even 67.4 %[4]. Especially road transport is nowadays almost solely dependent of the supply of mineral oil. EVs could contribute a lot to reduce this dependency.

2.2   Exhaust Emissions and Noise

Despite great improvements in the last decades, toxic exhaust emissions still are important negative by-products of road transport as well as noise. Especially in urban settings combustion engines of cars and trucks can add to the exceedance of air-quality limits, namely concerning particulate matter (PM) and nitrogen oxides (NO$_x$). These emissions are still in the focus when striving for better air quality in European cities (cf. discussion about low-emission zones in Germany, Italy or other European countries).

BEVs promise great improvements in this area because as a matter of principal they do not cause exhaust emissions locally. The same is true for PHEV while running on their electrical power train. For the cumulative effect on air-quality the emissions of electricity generating plants needs to be considered, of course. Though in large-scale plants elaborate emission control is possible so that good performance is achievable even when using fossil sources of energy. Further on electrical motors run very quietly and can reduce noise levels especially in urban transport. Though with increasing road speeds this advantage will diminish on average because noise from wheels and road surface will dominate, EVs will not contribute to the most annoying peaks of noise emissions since those are caused by combustion engines.[5]

2.3   Smart Grid, Vehicle-to-Grid

Besides its virtues for a better transport system EVs are regularly associated with benefits for the energy system. When charging is controlled properly EVs will be able to buffer electricity in times of weak demand (smart-grid) when generating plants otherwise could not be operated at economic load (especially at night

---

[3]   See Deutsche Physikalische Gesellschaft 2010, pp. 38–39.

[4]   Source: European Commission.

[5]   See Pehnt et al. 2007, p. 5.

times). Additionally EVs could be able to feed buffered power back into the electrical network in order to cover peak demand (vehicle-to-grid). Such contributions will be the more important the higher the share of solar and wind power will be as those energy sources cannot provide steady and foreseeable power supply. Nevertheless the potential of the vehicle-to-grid approach should not be overestimated regarding the still limited capacity of EV batteries even in optimistic scenarios of market development. At least in mid-term, benefits within the transport system will be the crucial arguments in favor of EVs.[6]

3   Technical Solutions

3.1   Electric Drive Vehicles

As mentioned the notion of electric vehicles can incorporate a range of technical approaches, from mild-hybrid to full battery-electric power-trains. Mild- and full-hybrids are only very limited implementations of electric drive technologies, though, because their battery capacity is usually quite low (up to 3 kWh) and they cannot be charged externally by using the electrical network. So they should be distinguished from more pure concepts that will be characterized as electric vehicles in this article. As such, plug-in hybrid electric vehicles (PHEV) are able to travel greater distances using only battery power (20 – 60 km or even more), carrying battery capacities from 6 to 15 kWh. Ultimately, battery electric vehicles (BEV) rely on electric power-train exclusively. Large battery capacities are needed to travel 100 km and more without recharging. In order to boost ranges of BEVs, either batteries that can be switched rapidly are considered or small combustion engines that will not drive the wheels of the car but charge the batteries (range-extender electric vehicles, REEV). While nickel metal hydride batteries as a well-known and broadly-used technology were sufficient for mild- and full-hybrid cars, plug-in hybrid electric vehicles and battery electric vehicles will need more advanced technologies like lithium ion high performance batteries.[7]

Table 1. Examples of electric drive vehicles in sale or in the testing stage.

Plug-in hybrid	Toyota Prius (in the testing stage)
Battery electric	Mitsubishi i-MiEV
Battery electric/range-extender	Opel Ampera
Battery electric/battery switch	Project Better Place

---

[6]   See Schill 2010, pp. 148–150.
[7]   See Schill 2010, pp. 140–143.

3.2   Other Alternative Drives and Energies

Besides EVs other technical solutions to reduce emissions and consumption of crude-oil are available:[8]

- First of all, there is still potential to optimize combustion engines. Fuel consumption of diesel cars is expected to drop further 8 % in the next years, consumption of petrol cars even up to 15 % by the use of new technologies. Hybridization – without stepping up to plug-in hybrid cars that can be charged externally – also is a mean to optimize conventional combustion engines by up to 25 % in fuel consumption (especially in urban driving conditions). Further developed combustion engines will not only be much more fuel-efficient, thus also reducing CO_2 emissions. Improved exhaust gas treatment can also minimize toxic emissions further (e.g. EURO 6 norm cars).

- Compressed natural gas (CNG) is a very promising alternative to diesel or petrol. It will not only reduce CO_2 emissions because of its lesser carbon content. There are also sources of natural gas that are much greater than those of crude-oil – opening up another stream of fossil power. Though of course also these fossil sources are exhaustible – only in a more distant future.

- Bio-fuels are produced from crops and can fuel combustion engines in different blends with fossil fuels or even purely (bioethanol, biodiesel, biogas). Depending on the production process bio-fuels can have a much smaller carbon-footprint than fossil fuels. Further on, bio-fuels are renewable sources of energy – independent of fossil sources. However, the processes used today to produce biodiesel or bioethanol are criticized to be little efficient and can create competition with the production of food. Next generation bio-fuels will mitigate these downsides. Concerning toxic emissions, combustion engines using bio-fuels will generally not perform better than ones using fossil fuels.

- Hydrogen can be used as a fuel for combustion engines as well as for fuel cells, producing electricity to drive an electric motor. Great expectations exist in hydrogen as it can be produced not only from fossil energy sources but also from renewable ones. Additionally only water vapor will emit when consuming hydrogen. However, a whole new network of filling stations will be needed to supply cars with hydrogen and to store hydrogen safely in cars is still elaborate and costly.

---

[8]   For further information see: ADAC 2009.

4   Barriers for Electric Drive Vehicles

4.1   Technical Restrictions

BEVs will bring specific restrictions that limit the group of potential buyers. At least for the next 10 or even 20 years BEVs as well as PHEV will have considerable disadvantages against ICE vehicles. Particularly the range of BEVs will be strictly limited to about 150 km without recharging the battery on the way.

Though daily average distances driven in Germany are only 27 km and about 85 % of all distances driven within one day are below 70 km, conventional ICE cars offer much greater flexibility than EVs. While those cars can be easily used for longer trips on weekends or for vacation this will not be possible with BEVs without relying on quick-switch batteries, range-extender or rapid recharging.[9] Only if car owners will use other means of transport (like rail) or car-sharing offers, BEVs will not be restricted to second car market for a very long time.

Consequently, *Baum et al.* expect that up to 2020 BEVs will spread as second cars only because of the limited range. Further they estimate that only drivers with their own private parking bay will be interested in BEVs because public charging infrastructure will develop slowly and will not be a feasible option for everyday use. Finally, batteries can only be recharged a limited number of times i.e. people travelling long distances altogether will not choose a BEV because of the high costs of replacing a used up battery. Considering all these restrictions a total share of 17.7 % of Germany's car stock is expected to be eligible to be replaced by BEVs in 2020.[10]

4.2   Costs

PHEVs acceptance will not be limited by operational range. Once the battery is exhausted the combustion engine can be used to drive on. For those cars in particular the total costs of running an EV compared with an ICE vehicle will be crucial. While operating costs will be favorable when using line current the investment of having a battery and an electrical motor next to a combustion engine will increase acquisition costs. Also attitude and habit will have a large impact when comparing costs: If battery range will usually be sufficient for the distances traveled the favorable energy costs of electricity will significantly reduce total costs. If otherwise the combustion engine will be needed often the additional acquisition costs for the hybrid system will probably not pay off.[11]

---

[9]   See Mehlin 2009, pp. 39–40.
[10]   See Baum et al. 2010, pp. 166–168.
[11]   See Mehlin 2009, p. 41.

Taxes and Costs of Energy

When studying costs of different energy sources to be used one has to account for taxation as a major component of energy prices. In the case of Germany – not atypical for European countries – fuels are taxed quite heavily at 65.5 Cent per liter petrol and 47 Cent per liter diesel compared with electricity.

Converted in taxation per unit of energy diesel (0.0479 €/kWh) is taxed more than twice as much as electricity (0.0205 €/kWh) and petrol (0.0772 €/kWh) almost fourfold.[12]

Should electricity used for cars be taxed as high as fuels its advantage in energy costs will vanish. Other way round, large market shares of EVs without changes in tax-rates will mean a distinct decrease in the fiscal revenues from energy taxes.

High performance batteries as needed for BEVs are still only available at very high costs. With an estimated manufacturer's price about 750 €/kWh (net of profits and taxes) a battery of 16–20 kWh will cause additional costs of 20˙000 € and more for buyers of such a car compared with a conventional ICE vehicle. Though a decline in prices at annual rates from 6–10 % is expected BEVs will carry a substantial burden of additional acquisition cost for many years to come.[13]

In order to calculate when total costs of an EV will break even with conventional cars, additional factors like electricity tariffs, prices of crude oil and manufacturing costs of the components needed (first of all for the different powertrains) have to be considered. With prices for crude oil between 100 and 130 $/barrel and electricity tariffs at 0.22 €/kWh *Baum et al.* expect for Germany in 2020 operating costs of 0.063 €/km for a BEV (about half the costs of a comparable ICE car) but acquisition cost that are 9˙600 – 15˙000 € above ICE cars.[14]

5   Perspectives of Electric Drive Vehicles

Though there is great interest in projections on the number of electric drive vehicles in the car fleet to be reached by 2020 or 2030 big uncertainty remains. The *International Energy Agency* describes in its BLUE MAP scenario what could be done to achieve the global goal of reducing annual CO_2 emission to half that of

---

[12] This is only a rough calculation, because only energy taxation (no additional charges on electricity) is considered and the taxation of primary energy (fuel) and secondary energy (electricity) is compared. Yet, the general direction holds even when integrating other charges and energy conversion efficiency into the calculation.

[13] See Baum et al. 2010, pp. 159–161.

[14] See Baum et al. 2010, pp. 162–165.

2005 levels by 2050. Strong policies and robust technological advances will be necessary to achieve this target, according to the *IEA*.[15] EVs play an important role in this scenario estimating global sales from BEVs and PHEVS about 7.2 million cars in 2020 annually rising to 33.9 million in 2030 and 101.3 million in 2050. Compared with the overall figures of new cars sold annually this would mean market shares of EVs from about 8 % in 2020 up to two thirds in 2050.[16]

Table 2. Projected number of BEVs in German car fleet by 2020 [million], depending on crude-oil price and annual decline in battery prices.

	130 $/bl	125 $/bl	115 $/bl	105 $/bl	100 $/bl
−10 % p.a.	1.414				
−9 % p.a.		1.054	0.824		
−8 % p.a.			0.583	0.377	
−6 % p.a.					0.104

Source: Baum et al. 2010, p. 180

Baum et al. project the number of BEVs in the German car fleet in 2020 in six different scenarios depending on crude-oil price and the annual decline in battery prices. Only in the two most favorable scenarios will the numbers exceed the Federal Government's target of 1 million cars by 2020.[17] In the "best-case scenario" a total number of EVs in the car fleet of 1.414 million would be achieved, corresponding with a market share of 12.61 % of all cars sold in 2020. In the "worst-case scenario" only 0.104 million EVs in the fleet would result, accounting for about 3 % market share in 2020.[18]

Germany's high level group on electric drive vehicles (Nationale Plattform Elektromobilität) predicts that without strong policies in favor of EVs the Federal Government's target of 1 million EVs will not be met by 2020 due to the disadvantages in costs. In that case only 450'000 cars sold by 2020 in Germany are expected.[19] Though, with a set of measures including changes to company car taxation, incentives to buy and privileged parking the target of 1 million is achievable, according to the group. Total sales are expected to consist of 50 % PHEV cars, 45 % BEV cars and 5 % utility vehicles.[20]

---

[15] See International Energy Agency 2009, p. 7.

[16] See International Energy Agency 2009, pp. 14–15.

[17] Though the 1 million target also accounts for PHEVs. Estimating an equal share of BEVs and PHEVs in car sales for the next years to come like other sources do, in four of the six scenarios the target would be met.

[18] See Baum et al. 2010, pp. 171–180. Schill 2010, pp. 144–145 quotes three additional projections for Germany.

[19] See Gemeinsame Geschäftsstelle Elektromobilität der Bundesregierung 2011, p. 43.

[20] See Gemeinsame Geschäftsstelle Elektromobilität der Bundesregierung 2011, pp. 31–32.

Though in spite of the often-cited benefits of electric drive vehicles there is skepticism about market development because of important uncertainties as to how quick the costs will decrease and what this will mean for acceptance.

However, *Schlager and Weider* point out that acceptance of innovative products might need to break with common attitudes towards established alternatives. First of all some people are interested in BEVs despite their limitations, people with openness to innovation, with a focus on environmental issues and desire to signal this. From evidence of field trials they further conclude that people often change their expectations when learning about new products like BEVs. Once drivers were able to use BEVs in their daily routine the attitude on range changed. Acceptance grew that the potentials offered by such cars are sufficient. Additionally, drivers appreciated the experienced advantages of BEVs like agility and low noise level. With reference to the success of the Toyota Prius hybrid car in the US also the important influence of emotional and symbolic aspects is stressed. Even with disadvantages about total costs the Prius found large acceptance as a car demonstrating the values and virtues of its owner as caring about the environment and about society.[21]

6 Conclusion

Electric drive vehicles have important potential to solve some of the challenges traffic faces today. Concerning their environmental footprint and the variety of energy resources usable, clearly e-mobility promises a major breakthrough in car technology. That said, the obstacles that prevented electrical drive to become the standard approach in cars still are yet to be solved. Even using new battery technologies developed for other applications – namely mobile devices in information and entertainment technologies – ranges of EVs are strictly limited and acquisition costs are high.

Optimistic market forecasts either rest on "best-case" assumptions (regarding the decrease of battery costs or large step-ups in prices for fossil energy resources) or ask for strong policies to advance EV sales. To answer if such policies as requested are sustainable in their fiscal requirements or even justified on the grounds of unemotional assessments of costs and benefits of technical alternatives is beyond the scope of this article.

Anyhow, looking into the mid-future until about 2020, EVs will not make a breakthrough contribution to reducing traffic's environmental- and climate-footprint – even according to the more optimistic forecasts. Other measures to be taken – e.g. further improvement in fuel efficiency of "conventional" combustion engine cars – have not become obsolete. Beyond 2020 though, EVs could reverse the market situation and become an important if not even dominant technology in

---

[21] See Schlager, Weider 2010.

road traffic – assumed that the now ongoing research and development will result in largely improved performance and costs of batteries.

References

ADAC (2009): Zukunftstechnologien – Was uns morgen antreiben wird. München.
Baum, Herbert; Dobberstein, Jan; Schuler, Bastian (2010): Nutzen-Kosten-Analyse der Elektromobilität. In *Zeitschrift für Verkehrswissenschaft* 81(3), pp. 153–196.
Deutsche Physikalische Gesellschaft (2010): Elektrizität: Schlüssel zu einem nachhaltigen und klimaverträglichen Energiesystem. (Studie der Deutschen Physikalischen Gesellschaft e.V.). Bad Honnef. Available online at http://www.dpg-physik.de/veroeffentlichung/ broschueren/studien/energie_2010.pdf, checked on 21/05/2011.
European Commission (2011): Roadmap to a Single European Transport Area – Towards a competitive and resource efficient transport system (White Paper). Brussels (COM(2011) 144 final). Available online at http://eur-lex.europa.eu/LexUriServ/LexUriServ.do?uri= COM: 2011:0144:FIN:EN:PDF, checked on 21/05/2011.
Gemeinsame Geschäftsstelle Elektromobilität der Bundesregierung (2011): Zweiter Bericht der Nationalen Plattform Elektromobilität. Berlin. Available online at http://www.bmwi.de/ BMWi/Navigation/Wirtschaft/Industrie/elektromobilitaet.html, checked on 21/05/2011.
International Energy Agency (2009): Technology Roadmap: Electric and Plug-in Hybrid Electric Vehicles. Available online at http://www.iea.org/Papers/2009/EV_PHEV_Roadmap.pdf, checked on 21/05/2011.
Mehlin, Markus (2009): Elektroautos für alle? DLR analysiert Nutzungspotentiale der Elektromobilität. In *DLR Nachrichten* (123, September 2009). Available online at http://www.dlr.de/vf/Portaldata/12/Resources/dokumente/DLR_N_123_Elektroautos_f uer_alle.pdf, checked on 21/05/2011.
Pehnt, Martin; Höpfner, Ulrich; Merten, Frank (2007): Elektromobilität und erneuerbare Energien. Heidelberg, Wuppertal. Available online at http://www.wupperinst.org/uploads/tx_wiprojekt/Energiebalance-AP5.pdf, checked on 21/05/2011.
Schill, Wolf-Peter (2010): Elektromobilität in Deutschland – Chancen, Barrieren und Auswirkungen auf das Elektrizitätssystem. In: Verkehr und Nachhaltigkeit (Vierteljahrshefte zur Wirtschaftsforschung, 2010/2), pp. 139–159.
Schlager, Katja; Weider, Marc (2010): Nutzerakzeptanz von Batterie-Elektrofahrzeugen. In *Zeitschrift für die gesamte Wertschöpfungskette Automobilwirtschaft (ZfAW)* (3), pp. 21–29.

Customer Needs and Attitudes Regarding Electrical Cars

Herbert Lechner

1 Introduction

The public discussion about the market introduction of electrical cars is strongly characterized by focussing on the supply-side especially on technical aspects like the maximum distance range of the battery or the models which will be available in the next years.

Due to the fact that there is only little data available from the demand-side the GfK Panel Services has done a representative consumer study amongst internet-

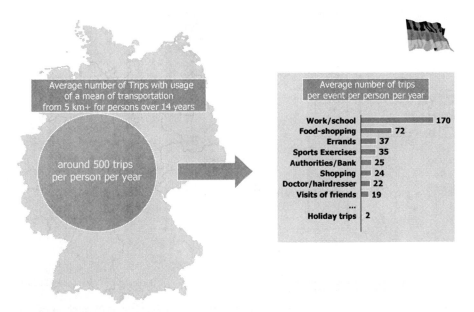

Fig. 1. Mobility behaviour in Germany: Germans are highly mobile for various occasions
Source: GfK

users aged 18years+ in Germany, Netherlands and UK for getting some empirical information about the potential customers and their attitudes and requirements towards these vehicles.

Having a look at the mobility behaviour in Germany it gets clear that actually most of the trips with the usage of any mean of transportation are in the local area: three out of four trips for work/school or for shopping are within a 20 km radius around the residence of the respective person.

Based on this mobility behaviour combined with a high ecological orientation in the German population it seems to be logical to introduce electrical vehicles as soon as possible. But do the potential customers really take into account what requirements these electrical cars have when purchasing them?

On being directly asked, only about 8–10 % of the consumers in Germany and UK really considered buying an electrical vehicle for their next car and respectively about one third of the consumers didn't yet know. In Netherlands even every third consumer intends to buy an electrical vehicle but not on the short run.

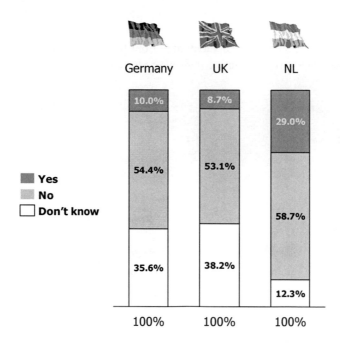

Basis: Adhoc-Online survey in 2010

Fig. 2. High general purchase intention rate of an electrical vehicle amongst consumers

Source: GfK

Generally the purchase intention rate increases in all countries with the educational level and the household income and reflects the necessary purchase power to afford a higher priced electrical vehicle compared to a fuel powered vehicle.

2 Potential Buyers

The market potential seems to be available in Germany for a short amount of time: about 40 % of those consumers with a purchase intention plan to buy within the next five years whilst the Dutch consumers mostly plan to purchase an electrical vehicle at the earliest from 2016 onwards or do not yet know exactly when they will buy such a vehicle.

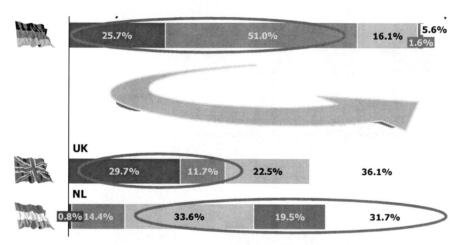

Fig. 3. Planned time period for purchase of an electrical vehicle mainly from 2013 on (only persons with purchase intention; Germany: n= 590; UK: n=358; NL: n=3133)

Source: GfK

The governmental support for the purchase of an electrical car amongst the three countries varies significantly: whilst in UK and NL the potential consumers get a direct bargain of € 5.000 – in Germany the incentive is only a reduction in tax on the maximum amount of € 2.000. Nevertheless more than half of potential German buyers are willing to accept a higher price.

Amongst those German buyers who accept a higher price more than 80 % are even willing to pay up to 2.000 € more than for a comparable fuel-powered car.

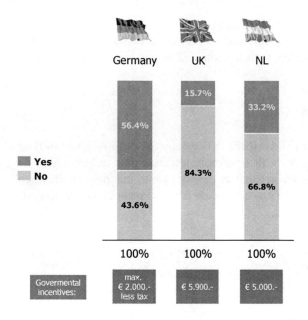

Basis: Adhoc-Online survey in 2010

Fig. 4. Willingness to pay a higher price achieves highest rate in Germany: every 2nd purchase intender accepts a higher price! (only persons with purchase intention)

Source: GfK

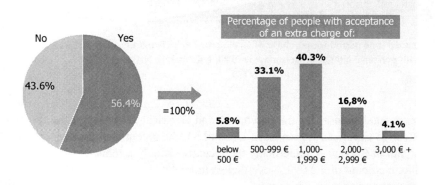

Basis: Adhoc-Online survey in August 2010 (n= 6,199 persons) aged 18 years + with Internet-Access

Fig. 5. Willingness to pay a higher price is up to 2.000 € amongst 80 % of purchase intenders in Germany

Source: GfK

3   Reasons of Non-buyers

About half of the car drivers in each of the three countries expresses that they actually do not intend to buy an electrical car for different reasons.

In Germany the main reason is clearly the limited range of the car due to the battery capacity. Their clear requirement is a driving distance of more than 300 km before the battery has to be recharged. All other reasons are of minor importance.

In UK and NL the limited range of the car is also an important barrier but not as dominant as in Germany due to the smaller size of their countries. Additional barriers are the high purchase costs and the low availability of charging stations.

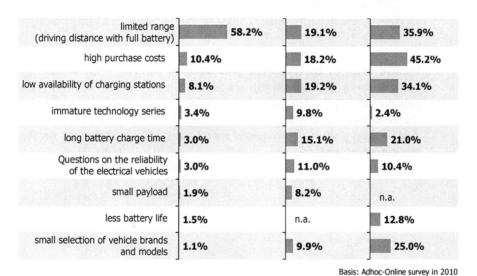

Basis: Adhoc-Online survey in 2010

Fig. 6. Main barriers for not planning to purchase an electrical vehicle are the assumed limited distance and the high purchase costs

Source: GfK

4   Conclusion

Based on this empirical information the conclusions are as follows:

- Consumer potential

  Basically there is a high willingness for sustainable mobility amongst consumers: already 9 % (UK) to 29 % (NL) of consumers are thinking about an electrical powered vehicle for their next car purchase.

- Target group of buyers

  The purchase intention increases with available purchase power and educational level in all countries. More than half of purchase intenders in Germany accept a higher price up to € 2.000,- for an electrical vehicle.

- Planned User behaviour

  Utlilisation as a second car is planned in household for downtown traffic within a distance up to 50 km a day.

- Customer requirements to driving range and speed

  A small car with a range up to 500 km and a maximum speed of 150 km/h is accepted by the majority. Battery charging from home is an absolute must.

- Barriers

  Limited range of battery, small penetration of charging stations and high purchase costs are the main barriers for the consumers in all countries. The potential of buyers will increase when the battery capacity improves and the investment in distribution of charging stations will be implemented. Additionally granted governmental incentives will be a catalyst for the further consumer demand.

Lightning Source UK Ltd.
Milton Keynes UK
UKOW03n1941231013

219638UK00001BA/55/P